❊❊❊❊❊❊❊❊❊❊

Planning for Seafood Freezing

Edward KOLBE
Donald KRAMER

MAB-60
2007

Alaska Sea Grant College Program
University of Alaska Fairbanks
Fairbanks, Alaska 99775-5040
(888) 789-0090
Fax (907) 474-6285
www.alaskaseagrant.org

Elmer E. Rasmuson Library Cataloging-in-Publication Data:

Kolbe, Edward.
Planning for seafood freezing / Edward Kolbe and Donald Kramer. – Fairbanks, Alaska : Alaska Sea Grant College Program, University of Alaska Fairbanks, 2007

126 p. : 51 ill. ; cm. (Alaska Sea Grant College Program, University of Alaska Fairbanks ; MAB-60)

Includes bibliographical references and index.

1. Frozen seafood—Preservation—Handbooks, manuals, etc. 2. Seafood—Preservation—Handbooks, manuals, etc. 3. Cold storage—Planning—Handbooks, manuals, etc. 4. Fishery management—Handbooks, manuals, etc. 5. Refrigeration and refrigeration machinery—Handbooks, manuals, etc. 6. Frozen fishery products—Handbooks, manuals, etc. I. Title. II. Kramer, Donald E. III. Series: Alaska Sea Grant College Program ; MAB-60.

SH336.F7 K65 2007

ISBN 1-56612-119-1

Credits

The work for this book was funded in part by the NOAA Office of Sea Grant, U.S. Department of Commerce, under grants NA76RG0476 (OSU), NA86RG0050 (UAF), and NA76RG0119 (UW); projects A/ESG-3 (OSU), A/151-01 (UAF), and A/FP-7 (UW), and by appropriations made by the Oregon, Alaska, and Washington state legislatures. Publishing is supported by grant NA06OAR4170013, project A/161-01.

Sea Grant is a unique partnership with public and private sectors, combining research, education, and technology transfer for public service. This national network of universities meets the changing environmental and economic needs of people in our coastal, ocean, and Great Lakes regions.

Editing by Sue Keller of Alaska Sea Grant. Layout by Cooper Publishing. Cover design by Dave Partee; text design by Lisa Valore. Cover photo © Patrick J. Endres/AlaskaPhotoGraphics.

Alaska Sea Grant College Program
University of Alaska Fairbanks
P.O. Box 755040
Fairbanks, Alaska 99775-5040
Toll free (888) 789-0090
(907) 474-6707 • fax (907) 474-6285
www.alaskaseagrant.org

✳✳✳✳✳✳✳✳✳✳

Table of Contents

Chapter 3 - Freezing Systems: Pulling out the Heat

Chapter 4 - System Selection and Layout

Chapter 5 - Scenarios

References

Appendix

Index

Preface

This manual is intended to serve as a guide for planning a seafood freezing operation. It addresses the physics of freezing, the selection of equipment and systems, and the important food science concepts that ultimately evaluate the process one would select. The format of this manual is similar to the publication *Planning Seafood Cold Storage* (Kolbe et al. 2006).

Our intended audience is plant managers and engineers; refrigeration contractors; seafood process planners, investors, and bankers; extension educators and advisors; and students in the field seeking applications information. We've based the content in part on our own experience. We've used the advice and information from industry. In large part, we've tapped the rich literature that remains from past specialists of the Torry Research Laboratory, the Canadian Federal Technology Labs, U.S. Sea Grant programs, the National Marine Fisheries Service, the Food and Agriculture Organization of the UN, and many others. Three overview reports are of particular value:

1. *Freezing technology, by P.O. Persson and G. Löndahl. Chapter 2. In: C.P. Mallett (ed.), Frozen food technology. Blackie Academic and Professional, London, 1993.*

2. *Planning and engineering data. 3. Fish freezing, by J. Graham. FAO Fisheries Circular No. 771. 1984. Available online at www.fao.org/DOCREP/003/R1076E/R1076E00.htm.*

3. *Freezing and refrigerated storage in fisheries, by W.A. Johnston, F.J. Nicholson, A. Roger, and G.D. Stroud. FAO Fisheries Technical Paper 340. 1994. 143 pp. Available online at www.fao.org/DOCREP/003/V3630E/V3630E00.htm.*

The sections of this book covering equipment and facilities explore options and give information that will help to pose the right questions so the reader can make good decisions. These must result from a knowledge of the seafood and the retention of its quality. The authors and publisher do not necessarily endorse the equipment described. This manual is not for design purposes, an area best left to industry specialists. Those wishing to pursue more details of modeling and engineering design can refer to a number of excellent chapters or books on these topics. Among them:

1. *Prediction of freezing time and design of food freezers, by D.J. Cleland and K.J. Valentas. Chapter 3. In: K.J. Valentas, E. Rotstein, and R.P. Singh (eds.), Handbook of food engineering practice. CRC Press, Boca Raton, 1997.*

2. *Food freezing, by D.R. Heldman. Chapter 6. In: D.R. Heldman and D.B. Lund (eds.), Handbook of food engineering. Marcel Dekker, New York, 1992.*

3. *Refrigeration handbook. American Society of Heating, Refrigerating, and Air Conditioning Engineers (ASHRAE), Atlanta, 1994.*

4. *Industrial refrigeration handbook, by W.F. Stoecker. McGraw Hill, New York, 1998.*

5. *Refrigeration on fishing vessels, by J.H. Merritt. Fishing News Books Ltd., Farnham, England, 1978.*

6. *Developments in food freezing, by R.P. Singh and J.D. Mannapperuma. Chapter 11. In: H.G. Schwartzberg and M.A. Rao (eds.), Biotechnology and food process engineering. Marcel Dekker, New York, 1990.*

Vendors and designers in the field can provide the ultimate recommendations for sizing and selection of specific equipment.

We have chosen to use the English system of units throughout the manual. It is a little awkward to do so, because the engineering and scientific world outside the United States has long ago moved to the SI (System Internationale) or metric system. The United States is not expected to follow any time soon. The Appendix gives a list of terms and their definitions, along with conversions between English and SI units.

❄❄❄❄❄❄❄❄❄❄

Acknowledgments

The authors acknowledge funding support for this project from the Alaska Sea Grant College Programs and Oregon Sea Grant College Program. Washington Sea Grant also contributed to the start-up effort. Thanks for help from the following:

Guenther Elfert, Gunthela, Inc.
Stuart Lindsey and Bob Taylor, BOC Gases
Ward Ristau and Randy Cieloha, Permacold Refrigeration
Greg Sangster, Integrated Marine Systems
Mike Williams, Wescold

✳✳✳✳✳✳✳✳✳✳✳✳✳✳✳✳✳✳✳✳✳✳✳✳✳✳✳✳✳✳✳✳

Author Biographies

Edward KOLBE
Oregon Sea Grant Extension, Oregon State University
307 Ballard Hall
Corvallis, OR 97331-3601
(541) 737-8692

Edward Kolbe is Sea Grant Regional Engineering Specialist (retired) and Professor Emeritus, Department of Bioengineering, Oregon State University. He recently held a joint appointment with Oregon Sea Grant and Alaska Sea Grant. For the last 30 years, he has conducted research and extension education programs to improve seafood processing, storage, and shipping. Kolbe's academic degrees are in the field of mechanical engineering.

Donald KRAMER
Marine Advisory Program, University of Alaska Fairbanks
1007 W. 3rd Ave., Suite 100
Anchorage, Alaska 99501
(907) 274-9695

Don Kramer is professor of seafood technology at the University of Alaska Fairbanks, School of Fisheries and Ocean Sciences. He also serves as a seafood specialist for the Alaska Sea Grant Marine Advisory Program. Prior to working at the University of Alaska, Kramer was a research scientist at Canada's Department of Fisheries and Oceans. His interests are in handling, processing, and storage of fish and shellfish. Kramer holds master's and doctorate degrees in biochemistry from the University of California at Davis.

❄❄❄❄❄❄❄❄❄

Chapter 1   Introduction

Before looking for answers, find out what the questions are.

PLANNING FOR A FREEZING SYSTEM

The role of a commercial freezer is to extract heat from a stream of product, lowering its temperature and converting most of its free moisture to ice. This needs to occur sufficiently fast so that the product will experience a minimum degradation of quality, the rate of freezing keeps pace with the production schedule, and upon exit, the average product temperature will roughly match the subsequent temperature of storage.

Before getting started on questions of equipment, power, and production rates, the paramount consideration must be the product. The well-worn message in any seafood processing literature is that "once the fish is harvested, you can't improve the quality." But what you can do is to significantly slow down the rate of quality deterioration by proper handling, freezing, and storage. Quality might be measured by many different terms: texture, flavor, odor, color, drip-loss upon thawing, cracking, gaping, moisture migration, and destruction of parasites, among other factors. For each species, these might be influenced in different ways by the physics of the freezing system—rate of freezing, final temperature, and exposure to air impingement, for example. And for each species, there may well be different market requirements that will influence your decisions. One example is albacore tuna. Freezing and storage requirements for tuna destined for the can will differ from requirements for tuna that will be marketed as a raw, ready-to-cook loin.

The first questions to address have to do with the product and the market, and Chapter 2 covers these topics both in general and in reference to species of interest. When your own product ques-tions have been addressed, you may then turn to vendors and contractors to plan or select the system. Their response will depend on some critical information that will affect freezing equipment options. The following list of critical information is adapted from recommendations of Graham (1974) and Johnston et al. (1994).

- *The anticipated assortment of fish (or other food) products to be frozen on this line.*

- *Possible future expansion need, extra production lines to be added.*

- *The shape, size, and packaging of each product.*

- *The target freezing time of each product.*

- *The product initial temperature.*

- *The intended cold storage temperature.*

- *Required daily or hourly throughput of each product, in pounds or tons.*

- *Normal freezer working day, in hours or numbers of shifts; schedule of workforce available to load and unload freezers.*

- *If a blast freezer:*
  Continuous or batch.
  The average air temperature required in the freezer section.
  The average design air speed required in the freezer section.

- *If a brine freezer:*
  Continuous or batch.
  Brine to be used, brine temperature, agitation velocity.

- *If a plate freezer:*
  Required plate temperature.

- *The position of the freezer on the factory floor, with a sketch showing its location in relation to other parts of the process.*

- *Maximum ceiling height and nature of the foundation available at the freezer location.*

- *Availability and specification of present electricity and water supplies.*

- *Reliability of electric supply and quality of water, and needs for backup sources.*

- *Maximum temperature in the surrounding room.*

- *Spare parts required and available, and reputation of vendor service.*

- *Availability of in-plant maintenance facilities and skilled labor.*

- *Equipment and operating costs and how they balance with anticipated product market value.*

- *The nature of the current refrigeration system available in the plant.*

- *If a mechanical (vapor-compression) system, the reserve refrigeration capacity currently available and the reserve power available.*

- *If cryogenic, the reserve capacity currently available and the local cryogen source and reliability.*

This list requires some explanation, and that is essentially the text that follows. One needs to understand the physics of freezing, how it is measured, and what factors control its rate. These are the topics of Chapter 1. If the freezing times are to be very short (minutes to complete the freezing), the freezer unit doesn't need to be very large to process a certain rate of production (in pounds per hour). The **refrigeration capacity** for such a situation, however, may be quite large; capacity is the rate at which heat is to be removed, in Btu per hour or refrigeration tons. (Note: terms, units, and conversions are in the Appendix at the end of this book.)

Details of how the freezing process will likely affect quality and other properties of specific seafoods are the topics of Chapter 2. Chapter 3 describes the heat sink, or refrigeration system you intend to use or install. There are two types. One employs the evaporation of refrigerant fluid that circulates within a closed cycle; this is called a vapor compression or **mechanical** system. The other type uses the low-temperature evaporation of expendable fluids—typically liquid nitrogen or liquid carbon dioxide; this is the open-cycle, **cryogenic system**. For each is described specific

freezing equipment that one can install; a partial list of suppliers for that equipment appears in the Appendix. Chapter 4 gives some sizing/selection details and such other system considerations as power consumption/cost, energy conservation, and planning onboard freezing options. Finally, Chapter 5 presents four examples or "scenarios" of freezer selection and processing.

THE FREEZING PROCESS

What Happens during Conventional Freezing

Heat flows from hot to cold. A freezing product is in fact "hot" compared to the "cold" surroundings that collect the heat. As heat flows out of each thin layer of tissue within the product, the temperature in that layer will first fall quickly to a value just below the freezing point of water. It then hesitates for a time while the **latent heat of fusion** flows out of that thin layer, converting most of the water there to ice. With time, the now-frozen layer will cool further, eventually approaching the temperature of the cold surroundings.

Figure 1-1 shows a temperature profile inside a freezing block of washed fish mince (known as surimi). In this simulation, a temperature profile is shown after 40 minutes of freezing. The upper and lower regions of the block (depicted here on the right and left, respectively) have frozen. The center layer (of approximately 1 inch thickness) is still unfrozen and remains so, as the heat flows outward and the moisture at its edges slowly converts to ice. Note that the unfrozen layer is at 29 to 30°F, lower than the freshwater 32°F freezing point. This is because of dissolved salts and proteins in the water fraction of the fish. The pace of the actual freezing process is relatively slow. To reduce the temperature of a pound of the unfrozen surimi by 10°F, we must remove about 9 Btu of heat. To then freeze that pound of product (with little change in temperature), we must remove another 115 Btu of heat.

We can calculate these temperature profiles for a whole series of times throughout the freezing cycle of this block of surimi, clamped tightly between the upper and lower cold plates of a horizontal plate freezer. Figure 1-2 shows these freezing profiles for a standard block thickness of 2.25 inches, wrapped in a single layer of polyethylene packaging, and in reasonably good contact with the top and bottom plates. In this example,

Figure 1-1. Simulated temperature profile inside a block of fish, 40 minutes into the freezing cycle. Initial block temperature was 50°F. External refrigerant temperature is –31°F.

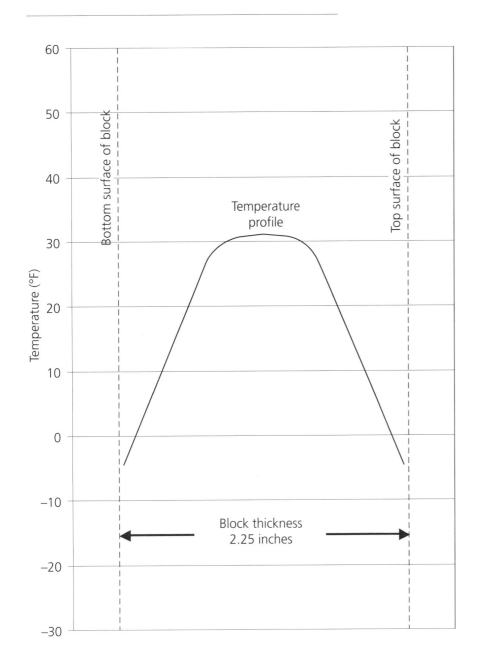

Figure 1-2. Simulated temperature profiles at 10-minute intervals within a block of surimi wrapped in a polyethylene bag and placed in a horizontal plate freezer. Assumed good contact between block and plates. Block thickness = 2.25 inches; initial temperature = 50°F; temperature of the freezer plates = –31°F.

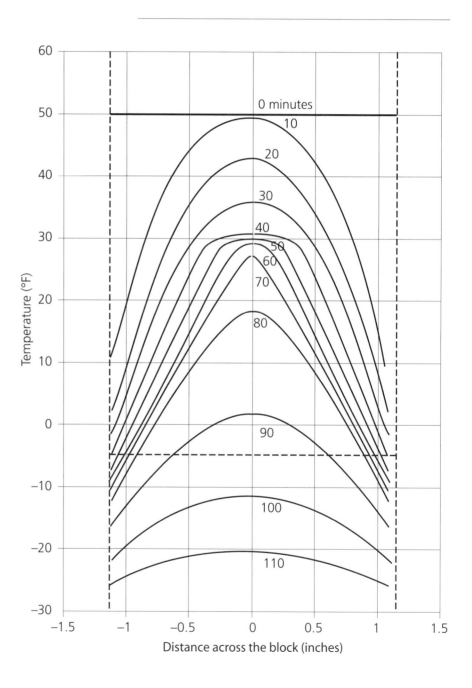

the initial uniform temperature at 0 minutes is 50°F, shown as a horizontal line near the top. The vertical dashed lines represent the upper and lower surfaces of the block. The horizontal dotted line represents the average block temperature after 90 minutes of freezing. Such an average would result if the block were suddenly removed from the freezer at 90 minutes and internal temperatures were allowed to equilibrate. Note that between about 40 and 60 minutes, the block's center temperature doesn't change much. This is the **thermal arrest zone**, which is characteristic of all freezing products. When we plot the center temperature with time (Figure 1-3), this zone appears as a plateau—a leveled-off section in the temperature history for the last point to freeze. Curves for two other loca-

tions show that off-center sites do not display this characteristic plateau.

These curves show the **freezing time** for this product—the time between placement of the block in the freezer, and the time at which the temperature at the core (or some average) reaches some desired value. For this example, it takes about 90 minutes to bring the core to 0°F. The process manager will need to know freezing times of various products and what controls them, as he plans the freezing operation and target production rate. In all cases with seafoods, the rate of freezing is important to ensure quality, although as we'll see, a "rapid rate" is a relative term.

What happens when fish tissue freezes? As heat flows away to a lower temperature region, the

Figure 1-3. Simulated temperature vs. time for three positions within the surimi block of Figure 1-2.

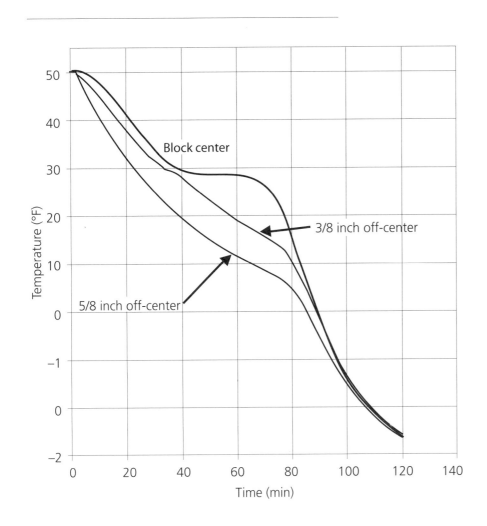

tissue first cools to a temperature at which a few ice crystals will be in equilibrium with the unfrozen fluid. As noted above, this will not be 32°F, as it is for tap water, but some lower temperature because of the various salts, proteins, and other molecules dissolved in that water. The initial freezing point of a fish (T_F) depends somewhat on its moisture content and is typically on the order of 28 to 30°F. A partial list of seafood and brine freezing points is in Table 1-1.

Ice crystals will begin to form at the site of solids or small bubbles within the fluid. Once formed, they grow into larger crystals as pure water in the fluid solidifies and combines with existing crystals. Because it is only the water molecules that solidify, the remaining "soup" becomes an even higher concentration of salts and proteins. This increasing solute concentration causes a continually decreasing temperature at which fluid and ice crystals are in equilibrium. Table 1-2 shows that this can continue to a very low temperature. The

Table 1-2. Unfrozen water fraction in haddock with moisture content of 83.6%

Temperature (°F)	Unfrozen water (%)
32	100
30.2	90.8
28.4	44.4
26.6	30.5
24.8	24.2
23	20.4
14	13.3
–4	9.4
–22	8
–40	7.8

Source: Riedel as reported by Fennema 1975

"frozen" haddock at –40°F still has 7.8% unfrozen water.

We would like to have the water freeze in the form of lots of small crystals, because that is ideal for maintaining the fresh fish texture. If the fish temperature falls rapidly to the initial freezing point, there can occur an overshoot for a short time. In this event, the temperature is colder than the initial (equilibrium) freeze temperature, but no ice has yet begun to form. The amount of **supercooling** is something that drives the rate of **nucleation**. The faster the temperature falls, the lower the supercooling temperature and so the higher the rate of nucleation or formation of new ice crystals. The rate of growth of these crystals with continued removal of heat (i.e., the **latent heat of fusion**) takes place more slowly. Thus anything that will speed up the rate of nucleation is a good thing.

So for fast freezing there are lots of small ice crystals; for slow freezing there are fewer and larger ice crystals. The terms "fast" and "slow" when referred to freezing really depend on the seafood product. Garthwaite (1997) reports that "quick freezing," as defined in the United Kingdom, is to reduce fish temperature from 32 to 23°F in 2 hours. He points out, however, that this isn't quick enough for products such as tails of scampi. Jul (1984) downplays the influence of freezing rates on quality. He expresses freezing rate in terms of "inches frozen per hour" and recommends that rates stay above $3/16$ inch per hour for most fish, and

Table 1-1. Initial freezing points (T_F) for seafoods and brines

Product	Moisture content (%)	Initial freezing point (°F)
Cod	82	30.4
	72	28.0
Crab	80.2	28.4
Hake	82	28.0
Halibut	75	28.0
Mackerel	57	28.0
Octopus	85–77	30.7–28.9
Oysters	87	27.0
Pollock	79	28.0
Salmon	67	28.0
Shrimp	70.8	28.0
Squid	84–79	30.9–29.0
Surimi	80	29.3
Tuna	70	28.0
NaCl brine	99	30.9
	95	25.5
	90	20.1
Sucrose brine	95	31.5
	90	30.9
	85	30.2

Source: Rahman 1995 and others

about double that for some shellfish such as shrimp and mussels.

In general commercial fish freezing terms, "fast freezing" will occur in hours; "slow freezing" will occur over days. Seafood technologists further stress that any quality advantage maintained as the result of rapid freezing can quickly disappear in a cold storage room that fluctuates in temperature or is not sufficiently cold (Fennema 1975, Merritt 1982, Cleland and Valentas 1997, Howgate 2003). Chapter 2 addresses this question in greater detail.

What is it that happens during slow freezing that can downgrade quality? There are several factors, many of which relate to the excessive time the product spends in the zone of partial freezing, between about 32 and 23°F. We could call this the **red zone**.

When the drop in temperature is rapid, the result is a high rate of ice crystal formation that occurs throughout the fish tissue—both inside and between the muscle cells. However, when freezing is slow, new crystals will tend to form first between muscle cells (a process called "extracellular ice formation" by Fennema 1975). With a slow temperature drop, the rate of ice crystal growth exceeds that of new crystal formation (Fennema 1975), and the extracellular crystals begin to grow. Where does the water come from to grow these crystals? It is pulled out through the muscle cell walls, leaving the cells partially dehydrated. The micro-image of Love (1966) shows the results of this process in cod muscle, Figure 1-4.

This slow freezing and the resulting large extracellular ice crystals present a number of problems affecting the quality of this fish:

Dehydration

When the fish is later thawed, the melting large extracellular crystals (Figure 1-4) become free water, most of which we'd hope to see permeate back into the muscle cells where it came from. This doesn't happen. Instead, it becomes **drip loss**, leaving a drier, tougher, less-tasty fish muscle.

Figure 1-4. Effect of freezing rate on the location of ice crystals in post-rigor cod muscle. (a) Unfrozen, (b) Rapidly frozen, (c) Slowly frozen.

(from Love 1966)

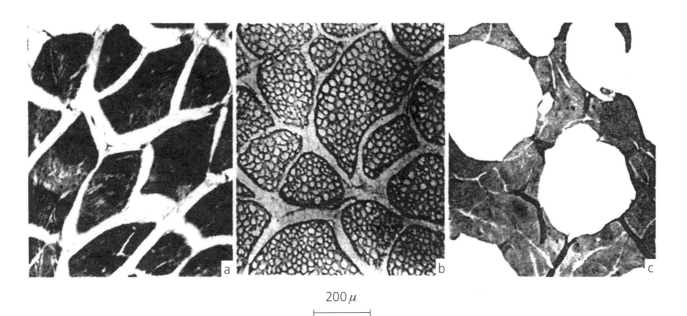

200 μ

Increased enzyme activity

In the red zone, there is an actual increase in the rate of chemical activity. As ice crystals form, the removal of water from solution concentrates the remaining salts, proteins, and enzymes into a soup that is still at a relatively high temperature, >23°F in this example. So there is still plenty of water left to support reactions. In the haddock example of Table 1-2, over 20% of the water is still unfrozen at 23°F. It is the enzymes in solution at a temperature that is still relatively high, that cause a number of quality problems that Chapter 2 describes in more detail. The picture in Figure 1-5 shows the increase in the rate of enzyme spoilage as the temperature falls through the red zone.

Denaturation

One result of enzyme activity in the red zone is protein denaturation. This means that muscle proteins have unraveled from their native, coiled-up state, and this permanently decreases their ability to hold water molecules. When the fish is thawed, the water, no longer bound to the muscle proteins, drains away as drip loss.

The actual effect of these and other consequences of slow freezing will depend on the product. For example surimi, which contains sugars and other **cryoprotectants**, would be far less affected than shrimp. Chapter 2 addresses more details of these impacts.

Alternative Processes

The above describes the usual process of food freezing as heat flows from the warm core out into the surrounding low-temperature environment. This freezing process, which can be completed within a few hours, is a typical one and can result in negligible damage to quality.

There are a few freezing processes that fall outside of this description and have some potential value. Most are under development and not yet in commercial use.

Figure 1-5. The effect of partial freezing on fish flesh for a hypothetical fish. Enzyme activity curves will be different for each species.

(from Doyle and Jensen 1988)

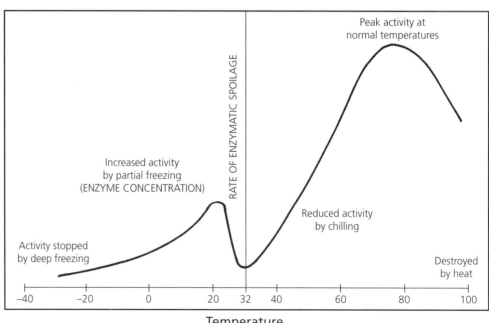

Planning for Seafood Freezing

Partial Freezing

Partial freezing refers to a process that brings the fish to some carefully controlled temperature below 30°F, at which only a fraction of the water—maybe 30-70%—is frozen. Other names used to describe these systems have been "super chilling," "supercooling," and "deep chilling." The fraction of frozen water reported as "best" varies with the species, application, and reporter. Those at the Torry Research Station in Scotland found best results with cod that were held at 28°F, when half the water was frozen (Waterman and Taylor 1967). Researchers in China reported commercial success with slightly warmer temperatures (–2°C, or 28.4°F), when 30% of the water was frozen (Ming 1982). A variety of species used in these experiments were common to the South China Sea.

Although partial freezing leaves the product in the enzyme-active red zone considered in the previous discussion to be undesired, there are a few possible benefits.

1. *The shelf life can be longer than that of iced fish landed unfrozen. British experience found that partially frozen cod from distant waters could last almost twice as long.*

2. *Ice crystal growth and damage to the fish tissue is perhaps less than it would be after more complete freezing, because a large percentage of the water remains unfrozen. (But it is critical to control temperature to maintain this low percentage of unfrozen water.)*

3. *Freezing fish at a temperature that is relatively high compared to that of conventional low temperature freezers means that the refrigeration equipment is more energy efficient and produces more capacity than it would at lower temperatures. (This characteristic of refrigeration systems is explained in a later section.)*

Naturally it takes a lot more refrigeration capacity—almost six times as much—to freeze a product (even just partially) than to chill it to above-freezing levels. Partial freezing also requires some very exact temperature control to ensure the best quality product.

Canadian fishermen once participated in a large-scale salmon experiment using partial freezing in weak brine onboard packers (tenders). Partial freezing sufficiently extended holding times to allow transport of Bristol Bay salmon to British Co-lumbia during the large sockeye run of 1980 when there was an anticipated surplus (Gibbard et al. 1982). So they sent several packers with well-designed and maintained refrigerated seawater (RSW) flooded tank systems, to buy and transport round, gillnet-caught salmon for canning.

Because of the long cruising time involved, it was necessary to find some means to extend holding time. They chose to partially freeze the product by adding salt so that the brine was about 6.5% salt by weight, then reducing the temperature to 26°F. (Pure seawater is about 3.5% salt.) They estimated that about 70% of the water in the salmon would be frozen at this temperature. The Technological Research Lab in Vancouver guided and monitored this large-scale experiment and considered the operation successful. Eleven vessels returned with about 4 million pounds of fish. Salmon unloaded at the cannery was up to 15 days old but still of acceptable quality. The experience highlighted several recommendations that include the following.

1. *The most significant factor affecting the quality of the loaded fish was handling prior to refrigeration. At that time (25 years ago), many of the fish were poorly handled, often unrefrigerated for up to 12 hours before loading onto the packer.*

2. *The RSW system must be good enough to bring temperatures to 27°F in 24 hours. Gibbard and Roach (1976) describe proper design of flooded onboard RSW systems.*

3. *The operator must make every effort to exclude air from the RSW mixture to minimize spoilage (oxidative rancidity) during transport.*

A series of four papers describing extensive partial freezing work prior to 1967 appears in an FAO volume on freezing (Kreuzer 1969). As a general statement, partial freezing is a difficult-to-control process that is not recommended by most seafood technologists.

Ultrasonics

Research over a number of years has shown that sound waves can help to increase the rate of nucleation, triggering the formation of small crystals, giving an increased texture quality (Nesvadba 2003). And because the sound energy goes right through the freezing product, the improved nucleation would occur as readily at the product center as at its surface. Recent researchers applied 25 kHz sound waves through transducers in a brine immersion tank held at 0°F (Li and Sun 2002a, Sun and Li 2003). Their experiments were to freeze raw potato sticks. They noted an increase in freezing rate and a significant decrease in tissue damage as acoustic power increased. We know of no commercial food freezers using this method, but work continues.

Pressure Shift Freezing

Pressure shift freezing, also called high pressure freezing, is another process that can force the rapid nucleation of tiny ice crystals throughout the food product being frozen. It is based on the characteristic of water that its freezing temperature will fall as pressure is exerted. The pressure shift freezing process is to first pressurize the packaged food product to around 2,000 atmospheres (about 30,000 psi) in a special liquid-filled cylinder, then chill it down to around 0°F. At that temperature and pressure, water is still in the liquid phase. When pressure is then suddenly released, the very large amount of supercooling (temperature below the initial freezing point) will cause a very high rate of nucleation. Continued withdrawal of heat then freezes the food product with a uniform distribution of small ice crystals (Denys et al. 1997, Martino et al. 1998, Schlimme 2001, Li and Sun 2002b, Fikiin 2003). The result can be a very high quality texture when the product is later thawed.

There are some problems. The high pressures have to be optimized—that is, selected to create a large super chill while avoiding damage to the muscle proteins; this could create a tough texture (Chevalier et al. 2000). The process is slow and the equipment expensive. In part due to these reasons, the process remains at the research stage; Europe and Japan may be making more progress toward commercialization.

High Pressure Forming

The high pressure forming process, observed several years ago, uses the same pressure-temperature properties of water to cause a uniform distribution of small ice crystals. Frozen portions were first tempered (warmed) to around 20-23°F, then pressed between two faces of a stainless steel mold. The pressure, exceeding 30,000 psi, caused the ice to melt (just as it does from the excessive pressure applied under a skate blade). As it melted under pressure, the portion assumed the shape of the mold. When pressure was then suddenly released, the portion's frozen water fraction was then made up of tiny ice crystals. Further freezing then took the product down to design storage temperature.

When glazed and held at a steady low storage temperature, a high quality, uniformly sized fish portion was the result. One observer recalled that there was some risk that the high pressure released and activated enzymes that could actually cause toughening in cold storage. It appeared promising, although expensive. It is not known if this process is currently used in commercial production.

FREEZING TIME: THE NEED TO KNOW

Why is it important that we know what the freezing time of our products will be? What things can we do to make it shorter? How do we find that out? Earlier sections have shown the importance of rapid freezing, or conversely, short freezing times. Other things also make short freezing times important. One is to reduce moisture loss. A rapid rate of freeze at the very beginning of the process is very important for products that are unpackaged. The quickly formed frozen crust will minimize moisture loss in air-blast freezers. Then a rapid finish of the freezing process will limit any further desiccation, or drying, and loss of water (that is, **weight**) to the high-velocity air swirling around the product. Note that such desiccation would take place by a process called sublimation—the transition of ice directly to vapor, while absorbing the heat of sublimation (see Appendix).

A second reason to determine and control freezing time is to minimize the time in the freezer. This is important for production reasons. Freezers are often the bottleneck in the process, and the operator wants to hustle the product through to keep pace with other parts of the line. The freezer is also an expensive piece of machinery. If it is possible to accelerate more product through without having to buy another (or larger) one, so much the better. Thus the pressure is on, for all of these reasons, to minimize time in the freezer.

It is also important, however, that the product be kept in the freezer long enough to ensure complete freezing. The consequences of taking products out too early, when they are not yet completely frozen, emerge in the cold store:

- *The requirement to continue freezing in the cold store room will overload the refrigeration machinery, and the room will begin to warm. Machinery there is not sized to freeze things; it is designed to maintain the low temperature and perhaps to slightly lower the temperature of already-frozen product that is moved into the room. Designers might typically assume new product temperature to be around 10°F warmer than the target storage temperature (Krack Corp. 1992, Piho 2000).*

- *Fluctuating temperature in the cold store room due to new, unfrozen product will cause fluctuating temperature on the surface of products that are already there. This leads to sublimation—ice transforming to water vapor directly, causing dried-out surface patches known as **freezer burn**. Although this won't happen in vacuum-packaged foods, packages with air spaces will fill with ice crystals as freezer burn proceeds.*

- *Incompletely frozen product eventually will finish freezing in the cold store. (In one particularly severe example with packages of pink shrimp, we've measured these delayed freezing times to be on the order of weeks.) The long freezing times that result will seriously affect quality.*

Thus, knowing how long the product must be held in the freezer is very important. The ability of a freezer to rapidly convert unfrozen products to frozen ones depends on two things. One is its **freezing capacity**; this relates to the refrigeration capacity available to that freezer. The other is **freezing time**, the time it will take the product to freeze.

Role of Freezing Capacity

Freezing capacity is the rate at which energy must be removed in the freezer to keep pace with the flow of product. It is the amount of heat that is to be taken out of each pound of product, multiplied by the production rate, in pounds per hour. **Rate of transferring energy** is the engineering definition of **power** (see Appendix), so freezing capacity relates directly to the required size of the refrigeration machinery—i.e., how much horsepower is involved. Engineers might describe capacity in units of Btu per hour, refrigeration tons, kilowatts (kW), and horsepower (HP). As Chapter 4 shows in more detail, refrigeration capacity must be adequate to remove heat from not only the freezing product,

but from a variety of other sources as well: fans, pumps, leakage, defrost, and others.

All of these heat loads make up the total freezing capacity that has to be matched by the capacity of the refrigeration. So freezing capacity and refrigeration capacity have to be balanced. Once you and the design engineers have arrived at machinery that you know will supply the refrigeration capacity needed, you must then consider the second and separate concept that will directly influence your production rate, and that is **freezing time**. Can the heat actually flow out of the product fast enough to meet a required time limit in the freezer?

Freezing Time: Influencing Factors and Calculations

Freezing time is the elapsed time between placement of the product in the freezer, and the time its average or core temperature reaches some desired final value. The final average temperature ought to be close to the expected storage conditions.

The curves of Figure 1-3 highlight several points of importance:

- *Recognizing the correct freezing time depends on an accurate measurement of temperature at the center, or **last point to freeze**.*

- *Presence of a **thermal arrest zone** is a characteristic of the last point to freeze and correct placement of a measurement probe. Only the **block center** curve of Figure 1-3 has the horizontal plateau, the thermal arrest zone.*

- *An incorrect freezing time will result if temperature sensors are not located at the last point to freeze.*

- *Prechilling would remove much of the heat above 29°F, decrease the overall freezing time, and increase the effective capacity of the freezer.*

Note that while **capacity** told us how big the compressors and heat exchangers had to be, **freezing time** will dictate how big the box, or freezer compartment, has to be. Consider the two freezers (the shaded boxes) of Figure 1-6. Both have the same capacity, expressed in product pounds per hour or refrigeration tons. But the example freezer at the top will freeze thin packages in 1 hour; in the bottom freezer, much thicker packages will take 3 hours to freeze. Although both would have roughly the same compressor size, the thick pack-

Figure 1-6. Two freezers with equal capacity, unequal product freeze time.

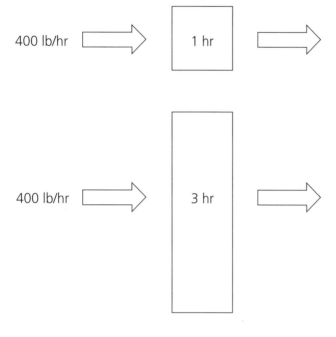

Freeze time affects optimum cabinet size

400 lb/hr → 1 hr →

400 lb/hr → 3 hr →

age freezer has to hold three times as much as the one freezing thin packages.

Plank's Equation

The time a product is left in a freezer is critical to production planning, cost, and potentially, quality. To control production flow, you must know freezing times for each situation, either by measuring directly or by calculating. Before discussing measurement, it is valuable to look first at one of the early and simple equations used to estimate freezing time, derived by R.Z. Plank in 1913 (Heldman and Singh 1981). Plank's equation, while not terribly accurate, is instructive because it includes the major influencing parameters:

$$T = \frac{\rho L}{(T_F - T_{Ambient})}\left[\frac{Pd}{U} + \frac{Rd^2}{k}\right]$$

where T = freezing time in minutes, or hours.

We don't really have much control over several of these parameters:

ρ = Product density (pounds per cubic foot, lb/ft³).

L = Heat of fusion (Btu per pound). This is the heat that must be removed to freeze the water in the product. (More on this in Chapter 4.)

k = Thermal conductivity of the frozen product (Btu per hr-ft-F). This is a measure of how quickly heat flows by conduction through the outer frozen layers. It is mostly a function of moisture (or ice) content but can be influenced as well by porosity and by direction of the muscle fibers.

T_F = Temperature at which product just begins to freeze (°F). A typical value for fish is 29°F, as indicated in Table 1-1.

P, R = Geometry terms that are fixed in value depending on the product's shape—whether it is a flat slab, a long cylinder, or a sphere.

There are three very significant parameters in this equation that we **can** control:

d = Product thickness or diameter (inches).

As the product freezes, heat is going to flow by conduction from the relatively warm center to the colder outer surface. As d increases, the heat flow path from the interior to the product's surface lengthens; the heat flow rate slows down, and freezing time increases. Sometimes not much can be done about the d in plate freezers, for example, where block thickness is fixed. However there may well be options with seafood products whose packages can be made thick or thin, stacked or not, in various kinds of freezers.

$T_{Ambient}$ = Ambient (or surrounding) temperature (°F).

The rate of heat flow from the product surface to the colder surroundings depends on the difference between those two temperatures, $(T_{Surface} - T_{Ambient})$. Double this difference and you double the rate of heat flow. $T_{Ambient}$ can represent temperature of the air-blast swirling around a seafood package. It can be the temperature of refrigerant as it flows and vaporizes inside the horizontal plates of a plate freezer. It can be vaporizing CO_2 "snow" inside a cryogenic cabinet, or sodium chloride brine spray in a fish hold. The lower (colder) this value, the higher the driving force $(T_{Surface} - T_{Ambient})$ pushing heat from the product, and the faster freezing takes place. However, as we will see in Chapter 3 (Figure 3-3), adjusting $T_{Ambient}$ downward will also reduce the capacity of the refrigeration machinery.

Although $T_{Ambient}$ is often assumed to be some constant value during the course of the freezing cycle, it will in fact typically increase somewhat at the beginning. Product is warm, the rate of heat flowing into the refrigeration system can briefly exceed its capacity, and refrigeration suction temperature (which relates to $T_{Ambient}$) initially rises. It then reaches a value at which the refrigeration capacity increases enough (Figure 3-3) to match the heat flow from the warm product. Later, as freezing proceeds and heat flow rate decreases, the saturated suction temperature falls back to its set point. In plate freezers, this happens more dramatically in the smaller, dry expansion (typically freon or halocarbon) systems than in the large liquid overfeed ammonia systems where it may not be noticeable at all.

Other things can locally affect $T_{Ambient}$ and thus freezing time. In a blast freezer, individual packages shielded from the blast may experience a local pocket of ambient air that is warmer than the air next to its neighboring packages. Oil clogging a flow passage inside a freezer plate would result in a local "hot spot" and a resulting increase in the adjacent block freezing time.

U = The heat transfer coefficient (Btu per hr-ft²-F)

Heat flow, Q (in Btu per hour), from the surface of a freezing product into the colder environment, can be described by the equation

$$Q = (U) \times (A) \times (T_{Surface} - T_{Ambient})$$

Where:
A = area of the exposed surface (ft²)
$T_{Surface} - T_{Ambient}$ = (defined above) (°F)
U = heat transfer coefficient (Btu/hr-ft²-°F)

U is a measure of how quickly heat is removed from the product's surface—through direct contact of the product with a freezer plate, or by the motion of a surrounding air, gas, or brine over the outside surface. U can have an influence on freezing time that is greater than that of the other parameters, but it is the most difficult-to-predict parameter in the freeze-time model. Its value usually results from a combination of surface resistances.

For plate freezing, it depends not only on the turbulence of the refrigerant flowing inside the plates, but also on the contact between block and plate, which in turn depends on pressure. The value of U, and so the rate of freezing, can be diminished by

- *A layer of frost on the plates.*

- *Layers of packaging material such as polyurethane and cardboard.*

- *Air voids at the product surface due to bunched-up packaging, an incompletely filled block, product geometry, or warped plates.*

- *Decreasing flow rate of boiling refrigerant inside the plate as freezing progresses and refrigerant demand falls. This is a minimal effect for large flooded plate freezing systems, but it can be quite significant for small dry-expansion plate freezers, as shown by experimental data of Figure 1-7.*

For air-blast or brine freezers, the value of U depends primarily on the local velocity. Brine usually gives a higher value than does air. U in these circumstances can also be diminished by resistances caused by packaging layers, air gaps due to packaging or geometry of the product, shading of fluid flow velocities that create a local dead spot, or poor freezer air circulation patterns. An example result of the last effect appears in Figure 1-8. Headed-and-gutted 6.5 pound salmon were placed on racks within a blast freezer. A diagram of a blast freezer appears in Figure 3-6. Because of poor balance of airflow (low at the top, high at the bottom), freezing times on the top shelves exceeded those on the bottom shelves by 4 hours (Kolbe and Cooper 1989).

The value of U in commercial freezers can be quite uncertain, and engineers must judiciously select, derive, or measure realistic values for prediction and design. Table 1-3 represents one set of numbers—generally on the low side.

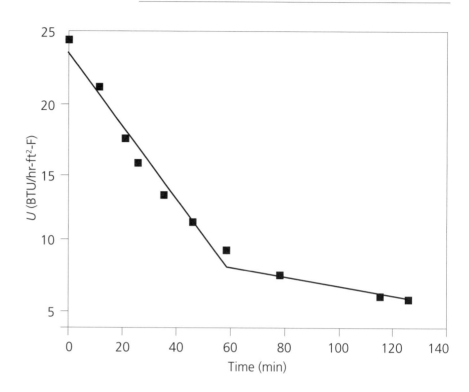

Table 1-3. Heat transfer coefficients, *U*

Condition	Coefficient, *U* (Btu/hr-ft²-F)
Naturally circulating air	1
Air-blast	4
Plate contact freezer	10
Slowly circulating brine	10
Rapidly circulating brine	15
Liquid nitrogen	
Low side of horizontal plate where gas blanket forms	30
Upper side of horizontal plate	75
Boiling water	100

Source: Heldman and Singh 1981

This is just one set of values. In fact, one can find far wider ranges of values measured for particular circumstances, as reported for various investigators by Cleland and Valentas (1997). For example, *U* for naturally circulating air (free convection) is 1-2 Btu per hr-ft²-F; *U* for air-blast freezers is 2-12; *U* for brine freezing is 50-90; *U* for plate freezing is 9-90 Btu per hr-ft²-F. An additional uncertainty is that the use of packaging materials can reduce the value of these coefficients to various degrees.

The influence of this range on calculated freezing time can be dramatic. Consider the example of a 1-inch-thick vacuum-packaged surimi seafood product. Say you want to freeze it from 60°F to a core temperature of –10°F in an air-blast freezer. Figure 1-9 is from a mathematical model that shows how freezing time varies with air velocity (Cleland and Earle 1984). Note that we assume, for this example, a representative commercial freezer in which the air-blast temperature is –10°F at the beginning of the freezing cycle, then falls rapidly to –25°F.

Figure 1-8. Measured variation of freezing times at varying shelf distance from the floor in a batch blast freezer. Product is 6.5 lb H&G salmon.

(from Kolbe and Cooper 1989)

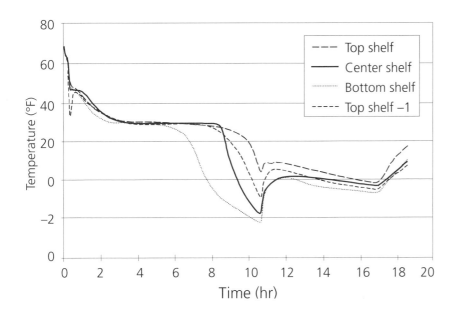

As velocity increases toward 20 feet per second, freezing time falls from about 2.5 hours and begins to level off at 1.25. Note that it is often a poor idea to increase air velocity much past 16 feet per second or so, because the more powerful fans begin to contribute a substantial heat load to the system (Graham 1984).

How can we shorten this freezing time? If we adjusted the freezer operating conditions and loading rate so that freezer air temperature remained steady at –29°F, prechilled the product so it entered at about 39° instead of 60°F, then removed it from the freezer when the core reached 10°F instead of –10, the resulting freezing time would fall to about 45 minutes. If we could then find a brine freezer having an agitated brine temperature of –40°F (there have been some so advertised), we might get the freezing time for this 1-inch package down to around 15 minutes. Thus it is possible to expect a freezing time range from 2.5 hours down to 15 minutes for this product, by manipulating external conditions, primarily those that affect the value of U.

Freezing Time Models

Knowledge of freezing time is important for the operation of freezers, management of production, and the quality, yield, and value of your product. How do you find out what it will be? There are two ways. Calculate it, or measure it. Although Plank's equation (described above) is a bit crude, many mathematical models have been more recently devised to simulate freezing (Cleland and Valentas 1997). Some of these are available in the form of relatively user-friendly computer programs. One program, developed at Oregon State University, used a complex modification of the closed-form Plank's equation to accurately describe freezing times (Cleland and Earle 1984). This older (DOS-based) program enabled users to select conditions covering a range of freezers and seafood product types, although it is not usually applicable to cryogenic freezing. A second computer tool, developed by Mannapperuma and Singh (1988, 1989) at the University of California at Davis, uses an enthalpy-based finite difference numerical model adapted to a user-friendly, Windows-based input format. It covers both freezing and thawing for a range

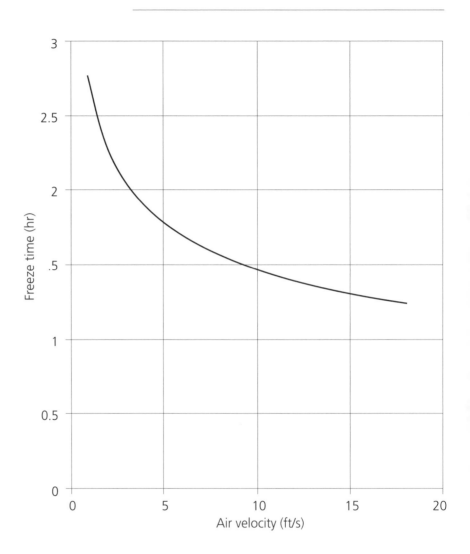

Figure 1-9. Influence of air velocity on freezing time. Product is a vacuum-packaged fish slab of 1-inch thickness. Initial temperature = 60°F; final core temperature = –10°F; air temperature range = –10 to –25°F.

of food products and allows one to visualize the internal temperature profiles. The program is distributed by the World Food Logistics Organization (2004).

These models enable you to play some "what-if" games with your situation. What if we made the package thinner, added a cardboard cover, increased air velocity, decreased initial temperature, and so on—essentially what we were doing with Figure 1-9. But it is very difficult for the operator to know, for example, the exact air velocity flowing over each package in a rack. (This relates to the difficult-to-predict value of *U*.) There are many uncertainties, and these will cause freezing time predictions to be off by typically 25% (Kolbe 1991). Sometimes ±25% is close enough, especially if the goal is to make comparisons. But for fine-tuning, the best way to know freezing time is to measure it.

Measuring Freezing Time

A temperature sensor in the core of a freezing product is the most direct way to measure freeze time. Figure 1-10 shows two fish mince packages in a small blast freezer, each starting at 55°F. If core temperature is to reach 0°F, freeze times of each is about 3.8 and 11.6 hours. In this case, a third sensor measured the air temperature, which was controlled to about –21°F ±6°F.

There are several approaches to freezing time measurement; each has some trade-offs that relate to accuracy, ease-of-use, cost, and risk of getting some bad results. The method chosen might also depend on whether you're trying to design a process or just trying to monitor final temperature to ensure quality in an existing process.

Drill and Probe

Probably the most common approach in the industry, particularly for frozen blocks, is to drill a hole and measure internal temperature with a probe (Graham 1977). The following paragraphs will describe the different probes that can be used. The procedure doesn't really measure freeze time, but it does tell the operator what the approximate core temperature is at the end of the freezing cycle. The procedure for drill-and-probe is this: Remove the frozen product from the freezer or cold store and immediately drill a deep (4 inches or more), small-diameter hole only slightly larger than the diameter of your temperature sensor. If the active part of the sensor is right at its tip, and if it contacts the bottom of the hole long enough for everything to equilibrate (maybe a minute or so), then you can expect to measure temperatures within 1°F of the true core temperature. However, if it is done wrong, errors can reach 35°F (Graham 1977, Johnston et al. 1994). Our own demonstrations with frozen surimi blocks have found that temperatures can easily vary from right-on to 10 or 12°F too high.

Equilibrium

In another, more-accurate post-freezing method, determine the average equilibrium temperature of products leaving a continuous freezer (Hilderbrand 2001). Load products into an insulated box with a temperature probe at its center. The warmer core and colder surface of individual products will come to an equilibrium after a short time. Wait until it does, then note the result.

Figure 1-10. Measured core temperatures of two package sizes in an air-blast freezer.

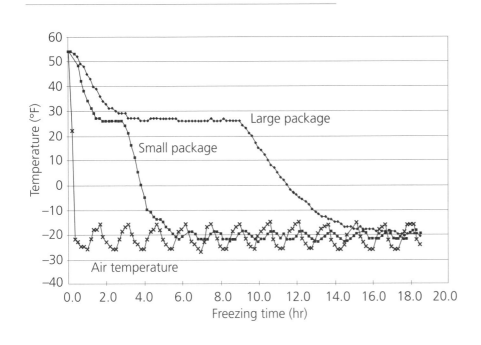

Table 1-4. Common thermocouples and ANSI color coding

Type	Positive wire		Negative wire		Plastic connector color
	Alloy	Insulation color	Alloy	Insulation color	
T	Copper	Blue	Constantan[2]	Red	Blue
J	Iron	White	Constantan[2]	Red	Black
K	Chromel[1]	Yellow	Alumel[3]	Red	Yellow

[1] Nickel-chromium alloy
[2] Copper-nickel alloy
[3] Nickel-aluminum-silicon alloy
Source: ASTM 1981, Omega Engineering 1998

Thermocouples

Clearly one of the best and most accurate methods of measuring freezing progress and time is with the use of thermocouples. A thermocouple is formed when two different metal alloy wires are welded (or soldered) together. The temperature of the junction (compared electronically with a reference temperature) is uniquely related to an electrical voltage, and this is transformed in a meter directly into a temperature readout. Table 1-4 gives three common thermocouple types and the color-coded conventions for insulation and connectors.

In its simplest form, a thermocouple probe is made of two insulated wires, exposed and fused together at the very end, then poked into the product center prior to freezing. In stationary freezers and cold rooms, the wires can be led out to a hand-held meter—the wire length doesn't matter. Then an operator can read and record the temperatures by hand every 10 minutes, or at some interval appropriate to the process. If plotted, these temperatures would draw, in effect, a curve that appears similar to Figure 1-3. Once freezing is complete, the wire can be snipped off (assuming the product is to be discarded) and a new couple quickly made by re-twisting and soldering the two wires.

In some temperature sensors, the thermocouple wires are fixed inside a thin-walled stainless tube, which often has a handle on the opposite end. The tube serves as the probe that can be poked into a fish or package quite easily. Wires are protected from breakage, moisture, and corrosion. The thermocouple probe pictured in Figure 1-11 was fabricated specifically for measuring core temperatures of freezing fish blocks. Once freezing was complete,

Figure 1-11. Thermocouple probe made of stainless steel tubing (³/₁₆ inch OD; 7¼ inch length). The right-angled handle allows twist and pull.

(from Kolbe et al. 2004)

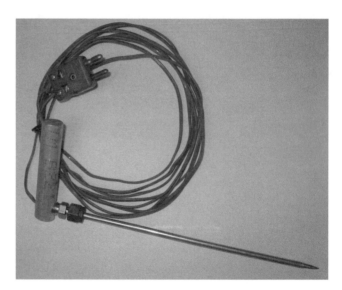

Figure 1-12. Bimetal thermometer with dual helical coil.

(from Claggett et al. 1982)

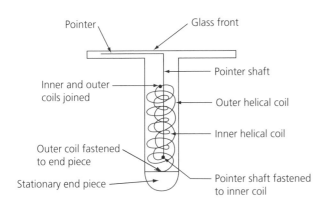

one then twisted the handle to break it free and pull it out.

Dial Thermometer

Such hollow-tube probes contain other types of temperature sensors as well. One hand-held type commonly used is the dial thermometer (Figure 1-12). As temperature changes at the tip, the bimetal coil tightens or loosens according to the different thermal expansion characteristics of the two metals. This action connects to a pointer at the free end that twists to indicate temperature on a dial. It is not hard to see how twisting, bending, bumping, or just age of this type of thermometer would lead to some serious errors.

Another hand-held type of dial thermometer uses a thermistor, a solid-state bead seated inside the tip of the probe. Its voltage signal is uniquely related to temperature. A small battery converts this signal to a digital read-out at the other end of the tube. The thermistor could also connect to a wire that then leads to a remote read-out or data-recording system. Thermistors are accurate and stable, but because of relatively high cost, are less disposable than thermocouples. They also may not operate correctly when the battery-driven instrumentation is exposed to very low temperatures.

Thermistor Data Loggers

The temperature measurement devices just described are all most suitable for stationary or real-time measurements. What do you do if a product is not stationary? For example, you may want to know the temperature of a fillet as it travels through a spiral freezer. Or you may want to follow the temperature of a container-load of frozen blocks delivered by truck or barge. Partly as a result of HACCP food safety programs, we have seen development of a large number of very small, easy-to-use data loggers that can sample, store, then report temperatures (among other things) at whatever rate you select. Some have the sensor element built into the body of the recorder. But if freezing time is to be measured, you'll need one with a remote sensor that can be poked into the center of a product. Some difficulties result if the connector between sensor and logger is not sealed, allowing moisture to condense onto the connector or electronics as the logger is removed from the cold room. In addition, some recorders (and batteries) cannot themselves operate when freezer temperatures get too low.

All temperature measuring devices can be wrong, due to damage, age, or other factors. You can account for these kinds of instrument errors by calibrating them from time to time. Errors can also result from the wrong measuring procedures. The Appendix describes a calibration method for measurements in the food freezing ranges, and various sources of measurement error.

SOME "WHAT-IFS"

It is impossible, or at least not cost-effective, to measure freezing times for each combination of packaging, geometry, product, and temperature schedule. This highlights the value of simulation models that can indicate the results of such combinations, as long as the user can systematically check results using selected experiments. The example of Figures 1-2 and 1-3 gives one such simulation that resulted from applying of a software package called PDEase (Macsyma Inc., Arlington, Massachusetts). This is a partial differential equation solver and is representative of models previously used and verified for seafood freezing (Wang and Kolbe 1994, Zhao et al. 1998). This section will further use this numerical model to investigate some what-if questions that may influence production rate.

Fish Blocks in a Horizontal Plate Freezer

The first set of examples considers a block of washed fish mince, or surimi, containing a cryoprotectant mix of 4% sucrose and 4% sorbitol. The thermal properties were defined by Wang and Kolbe (1990, 1991). With an 80% moisture content of the mince, this product will respond in a similar fashion with other fish blocks or packages that are densely packed.

Block Thickness Effects

The example of Figures 1-2 and 1-3 show freezing time for a 2¼-inch block in a plate freezer operating under a given set of conditions. Figure 1-13 shows the results of this program for different block thicknesses and with other conditions held constant. The assumed final average temperature in all cases is about –4°F.

An approximate rule of thumb can be used when the heat transfer coefficient, U, is very large. In that case, the freezing time will be about proportional to the square of the thickness. Plank's equation (given earlier) will show this. For the case of Figure 1-13, the U value was low enough to throw this rule off a bit.

Cold Temperature Sink Effect

The simulation of Figure 1-14 shows how block freezing time is expected to vary with an assumed fixed refrigerant temperature inside the freezer plates. The lower temperatures and freezing times will effectively increase the pounds-per-hour that can be processed and frozen. The lower limit on the refrigerant temperature will be dictated by the refrigeration plant capacity that must always be sufficient to do the job. This also falls as temperature lowers.

Internal plate temperatures in small plate freezers may actually increase somewhat at the beginning of a freeze cycle, and this could extend the expected freezing time. Normally, for ammonia flooded or liquid-overfeed systems, this increase would be quite small (Takeko 1974). However, when the product capacity greatly exceeds the refrigeration capacity, as in an overloaded blast freezer for example, that suction temperature would rise appreciably as the rate of heat flow from the freezing product adjusts to match the rate of heat absorbed by the refrigeration machinery.

Heat Transfer Coefficient Effect

Figure 1-15 gives predicted freeze times for a range of external heat transfer coefficients, U. The same surimi block used in the calculations above is used to show how various combinations of external velocities, media, packaging—which all determine the value of U—affect freezing time. Note that freezing times are less and less influenced by U as U gets large. (A near-minimum freezing time of 55 minutes is reached when U is close to 90, not shown). As U increases, there will be some value at which the internal resistance to heat flow begins to "control" the process. That is, even if heat leaves the outer surface so readily that the surface temperature becomes almost equal to that of the surroundings, there will still remain an internal resistance to heat flow. And that is why freezing time never goes to zero. A complicating factor in any freezer is that conditions causing low U values,

Figure 1-13. **Predicted freezing time vs. fish block thickness. Product is surimi, 80% moisture content, 8% cryoprotectants, –31°F cold plate temperature, 50°F initial temperature, –4°F final average temperature. Block is wrapped in polyethylene bag with good contact with freezer plates.**

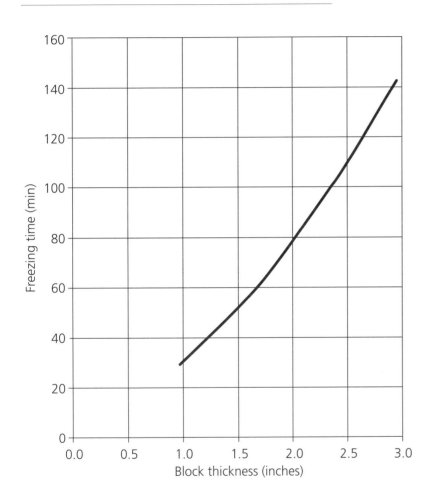

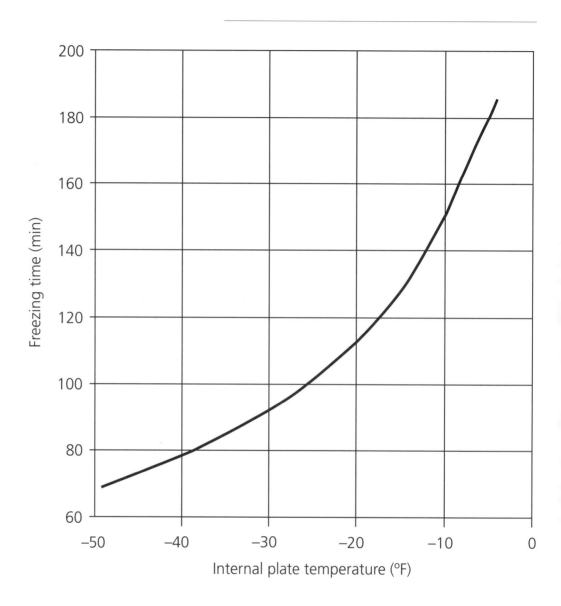

Figure 1-14. Predicted freezing time vs. freezer plate temperature. Block thickness is 2.25 inches. Other properties described in caption of Figure 1-13.

Figure 1-15. Predicted freezing time vs. heat transfer coefficient *U*. Surimi block thickness = 2.25 inches; other conditions as described with Figure 1-13.

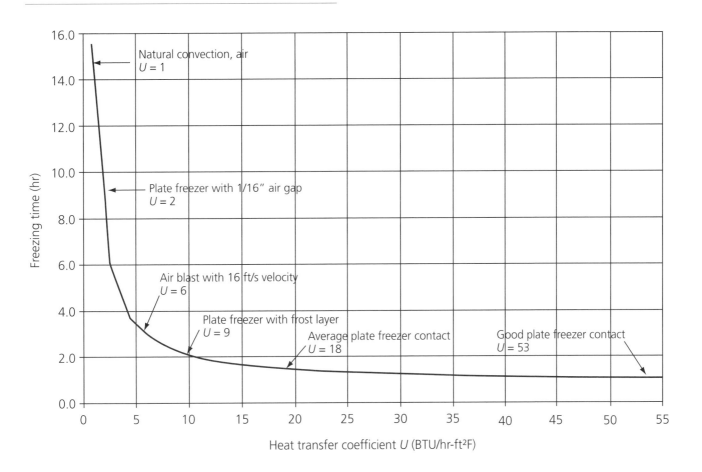

and so excessive freezing times, may not be uniform throughout the freezer. This can lead to some incompletely frozen product being shipped to the cold store.

Commonly, there will be different conditions of contact—and so different values of *U*—on different surfaces of the freezing product. In those cases, one cannot expect to find the symmetric freezing pattern seen in Figures 1-2 and 1-3. Instead, an off-centered freezing pattern will often lead to a longer freezing time. As an illustration, we will repeat the freezing simulation of Figures 1-2 and 1-3, still assuming a fixed plate temperature, $T_{Ambient}$. Figures 1-16 and 1-17 simulate the freezing block with the same good contact on the bottom

side. On the top side of the block, however, there is now a $^1/_{16}$-inch air gap. Such a condition could be caused by a warped pan, an underweight (and thinner) block, a heavy frost layer, bunched-up poly bag, or corrugated cardboard packaging. The results highlight several things. Figure 1-17 shows that it took essentially twice as long (180 minutes) to achieve the same average temperature of –4°F as did the balanced freezing block. The block freezes unevenly, as expected (Figure 1-16). The geometric center is not the last point to freeze. As shown by Figure 1-17, a temperature measured at the block's center would fail to display the plateau, or **thermal arrest zone** seen for the geometric center of the block frozen in Figure 1-3. It would in fact indicate to the operator that freezing was complete quite a while before it actually was.

Whole Albacore Tuna Frozen Onboard

West Coast commercial albacore trollers will commonly freeze onboard with trip lengths exceeding a week or so. There are two common systems. One is spray-brine, where cold sodium chloride brine is sprayed over the fish placed into the hold. The other is air-blast, where fish are hung or placed on racks as cold air is circulated with fans. Because albacore is a warmer-blooded fish than others such as salmon, the core temperature of the struggling fish landed on deck can approach 80°F. High quality requires that this heat be promptly removed. This can be accomplished either in a deck tank, or by quickly moving it to a below-deck freezer.

Simulation models allow one to anticipate how various factors will affect the chilling and freezing rates. Zhao et al. (1998) demonstrated a model, verified with experiments, that presents a good prediction. Because heat flows out through the oval-shaped body of the fish, the modelers' measurements also showed how these body dimensions (and heat removal characteristics) can be correlated with fish weight. The following examples of "what-ifs" for albacore tuna show the effects of onboard handling. Other examples are described by Kolbe et al. (2004b). The relationships shown can be used to anticipate how other fish chilling and freezing changes might occur.

Blast-Air Temperature Effects

Both the design and loading rate of an onboard blast freezer can alter the operating temperature in these freezers. Figure 1-18 predicts how freezing time of a 24-pound tuna will vary with air temperature. The air temperatures and velocity used (6 feet per second) are not ideal; they can be representative of small-boat freezers. In fact, recent markets for high-valued albacore products will push required freezing and storage temperatures to lower values, approaching –40°F.

Freezing Medium Effects

Under typical conditions, the cold air medium of a blast freezer and the cold brine medium of a spray brine system will freeze fish at different rates. The heat transfer coefficient, U, of a liquid is almost always greater than that of a gas. (For a spray-brine freezer, this requires a direct and uniform coverage of the sprayers—difficult to achieve on small boats.) On the other hand, we can control air-blast temperature to a far-lower value than sodium chloride brine, whose practical lower limit is about 0-5°F. Figure 1-19 shows that if each system were well-designed and operated, the freezing times could be roughly similar. The curves also predict the effect of fish size.

Prechilling Effect

There are three benefits of prechilling in an on-deck tank—either with slush ice or RSW (refrigerated seawater). First, it promptly begins removing heat from the fish, producing well-documented advantages to final quality and safety. Second, by removing a portion of the heat (on the order of 25%) in the deck tank, the effective refrigeration capacity is similarly increased. And third, you can reduce temperature fluctuations in the hold—something that is damaging to the quality of fish previously frozen and stored.

The model has been used first to show prechilling rates (Figure 1-20) in various media. Note that the chill rate for fish packed in ice would fall somewhere between that for slush ice, and that for still, cold air. Figure 1-21 then shows how this prechill can affect the freezing time.

Figure 1-16. Simulated temperature profiles at 10-minute intervals. The freezing fish block is surimi in a horizontal plate freezer. On one surface: good contact (U = 18 BTU/hr–ft²–F); on the opposite surface: a ¹/₁₆-inch air gap (U = 2 BTU/hr–ft²–F). Initial product temperature = 50°F. Freezer plate temperature = –31°F.

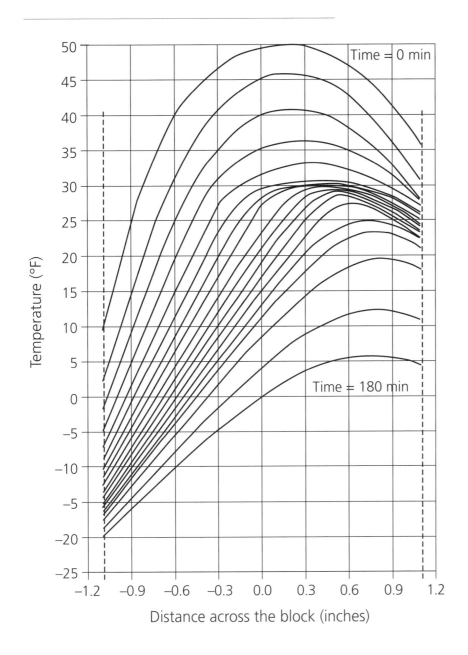

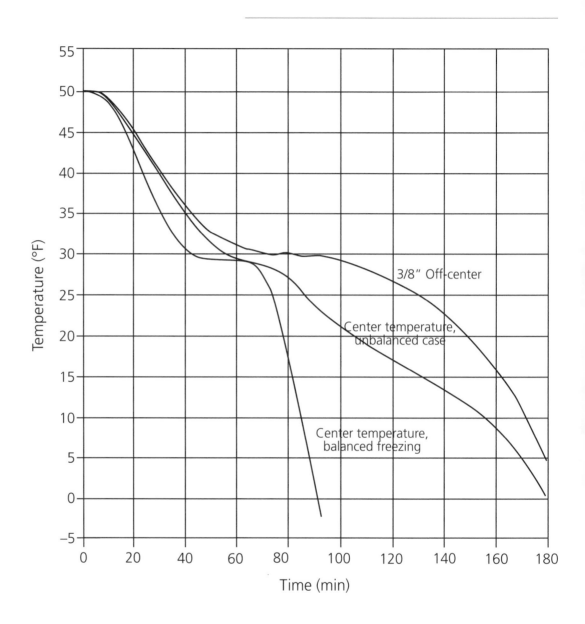

Figure 1-18. Predicted core temperature of 24 lb albacore tuna. Three blast freezer temperatures. Air velocity = 6 ft/s. Some on-deck prechilling assumed.

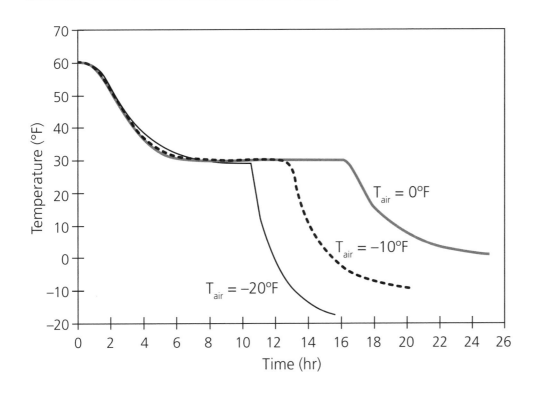

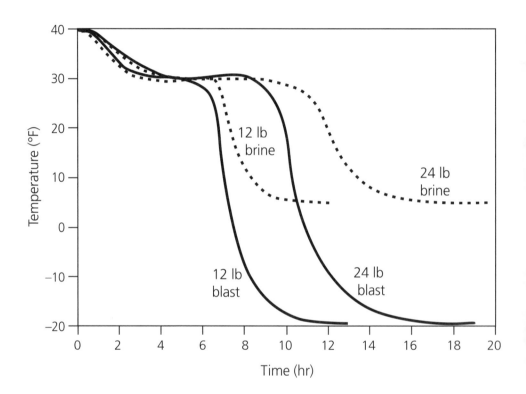

Figure 1-20. Predicted on-deck prechilling rate for a 24 lb albacore tuna.

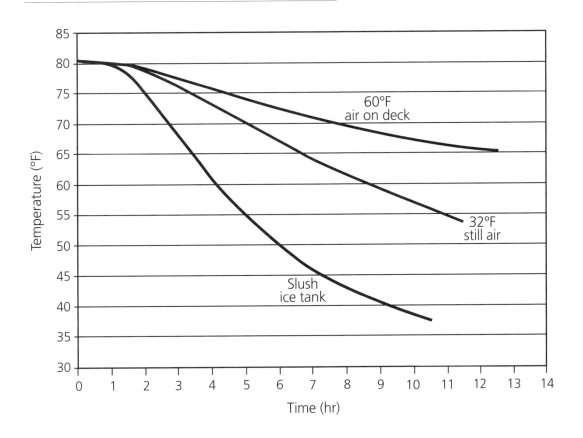

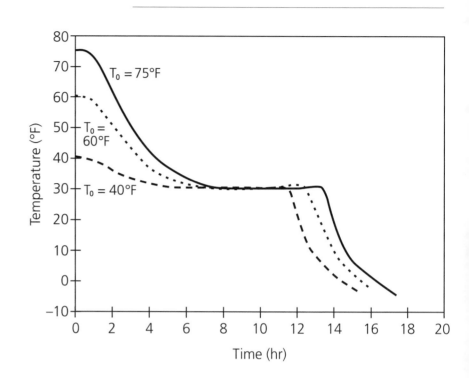

Planning for Seafood Freezing

Chapter 2 Freezing Effects on Fish and Other Seafoods

Freezing and frozen storage are almost universally accepted as the preservation method of choice. Because the nature of the product is not appreciably altered, compared with salting, drying, smoking, and canning, the use of freezing, cold store, and thawing yields a seafood product most like the fresh product.

The artificial freezing of seafood products began in the mid to late 1800s using ice and salt. The frozen product was glazed with water and then stored. Several patents were issued on this process. Many other patents were subsequently granted on freezing using eutectic ice. The use of ammonia refrigerating machines for freezing began in the late 1800s. Shipment of frozen seafood was developed in the early 1900s using dry ice.

Freezing functions as a preservative by

- *Reducing temperature, which lowers molecular activity.*

- *Lowering water activity, which lowers microbial growth.*

Only the very best freezing and frozen storage processes yield high quality. Otherwise you will have a fair to poor product. Air-blast and contact plate freezing systems can reduce the temperature of round fish at the warmest point to 0°F within 24 hours. For portions no thicker then 2 inches, the temperature at the warmest part can be reduced to 0°F in 2.5 hours. For large round fish more than 6 inches thick, it can take up to 72 hours to get the temperature at the warmest part of the fish down to 0°F. Use of a brine freezer with clean brine at 7°F or lower can reduce the temperature to 15°F at the warmest part of the fish.

The freezing step often gets the major emphasis. However, holding in the cold store and thawing must receive equal consideration in reducing qual-

ity loss. The most serious loss of quality is usually associated with these latter parts of the cold chain.

The main causes of deterioration during freezing and frozen storage are

- *Oxidation of lipids resulting in rancid odor and flavor.*

- *Toughening due to protein denaturation and aggregation.*

- *Discoloration largely due to oxidation reactions.*

- *Desiccation (freezer burn).*

PHYSICAL AND CHEMICAL CHANGES DURING FREEZING

Flavor and Odor Changes

Freezing and thawing, as well as frozen storage result in

- *Decrease in intensity of fresh odor and flavor for seafood frozen when very fresh.*

- *Rancid odors and flavors, especially in fish high in lipids.*

- *Enzymic catalyzed changes resulting in cold storage odors and flavors not due to lipid oxidation.*

Many fruits and vegetables have enzymes that can catalyze changes during cold storage, which lead to off-odors and off-flavors called cold-store odors and flavors. Blanching to inactivate these enzymes is used to control this problem in fruits and vegetables. With fish and shellfish, this is not such a serious problem.

When good freezing and thawing practices are used, there will be a very small effect on fish quality. However, a well-trained panel can tell the difference between unfrozen fillets and fillets that have been frozen and immediately thawed.

Table 2-1. Effect of freezing and storage on total bacterial counts of haddock

Time at which counts were made	Total bacterial count per gram		
	Fresh sample	Slightly stale sample	Stale sample
Just before freezing	25,000	500,000	12,000,000
Immediately after freezing	1,500	28,000	950,000
After 1 month at 0°F (−18°C)	900	16,000	430,000
After 6 months at 0°F (−18°C)	00	14,000	300,000
After 12 months at 0°F (−18°C)	600	11,000	270,000

Source: Licciardello 1990

Double freezing and thawing will result in more quality loss than single freezing and thawing, as there will be two passes through the critical zone (28 to 23°F) during freezing. However, the quality can be satisfactory if the freezing is not done too slowly.

Flavor and color changes result from the oxidation of oils and pigments. Acid and carbonyl compounds are formed that have unpleasant odors and flavors. Rancidity can develop very quickly after freezing, especially in fatty fish but also in lean fish. The effect in lean fish is often not blamed on rancidity but is simply called cold-store odors or flavors. It is important to be aware that development of rancidity is not simply dependent on the amount of oil present. A good example is that pink salmon lipid is less stable to oxidation than chinook salmon lipid, which has a larger amount of lipid. Sablefish is high in lipid but the lipid is more stable than would be expected based on the amount present.

In the past it was necessary to keep frozen fish separated from other foods due to odor transfer. Now with better packaging (resistant to passage of gases), there is no need to separate fish from other frozen foods.

Effect on Microorganisms

Bacterial activity is inhibited in frozen seafood since bacteria need free water. Some psychotropic bacteria can grow at below the freezing point, but the rate will be much lower than in unfrozen tissue. Growth of microorganisms is relatively common at 23°F down to 19°F. Growth is rare at 15°F, and below this, growth of microorganisms is largely inhibited and they become dormant.

Freezing and frozen storage have a lethal effect on some bacteria (Sikorski and Kolakowska 1990). Freezing can destroy as much as 50 to 90% of bacteria (see Table 2-1). During frozen storage there is a slow, steady die-off, the rate of which depends on storage temperature and bacterial species (Licciardello 1990). The lethal effect is highest between 25°F and 14°F rather than at lower temperatures. The most sensitive bacteria are gram negative bacteria vegetative cells, and the most resistant are spores and gram positive bacteria. Reduction can be as high as two orders of magnitude.

In the preceding paragraph we have mentioned the lethality of freezing and frozen storage (up to 90% or two orders of magnitude), but this is not always the case. A study on cod and ocean perch fillets that were frozen and thawed after no more than 5 weeks in storage showed very little change in total bacterial counts. However, there was a reduction in total counts after storage for 14 weeks at −13°F (Magnússon and Martinsdóttir 1995).

Upon thawing, the bacteria surviving freezing and frozen storage will grow and multiply. It has been reported that the surviving bacteria multiply faster due to changes brought about by the freezing and thawing. However, it has also been reported that thawed fish does not spoil faster than never-frozen fish. There could be several factors that account for these seemingly contradictory reports:

- *Bacterial flora.*

- *Prefreezing fish quality.*

- *Freezing rate.*

- *Storage conditions (temperature and fluctuations).*

- *Method of thawing.*

An excellent review article on freezing as a method of providing safe food was written by Archer (2004). He points out that more research is needed to understand how the resistance of human pathogens to the lethal effects of freezing is induced. Understanding this could make freezing a more reliable way to provide increased food safety.

Effect on Parasites

Internal parasites found in fish, shellfish, and aquatic mammals include the following:

Nematodes (roundworms)
  Anisakis spp.
  Contracaecum spp.
  Eustrongylides spp.
  Gnathosloma spp.
  Phocascaris spp.
  Phocanema (*Terranova*) spp.
  Pseudoterranova spp.
  Thynnascaris (*Hysterothylacium*) spp.
  Trichinella spiralis

Cestodes (tapeworms)
  Diphyllobothrium spp.

Trematodes (flukes)
  Heterophyes spp.
  Metagonimus spp.
  Opisthorchis spp.
  Paragonimus spp.
  Chlonorchis sinensis
  Nanophyetes salminicola

The life cycles for these parasites are often very complex. The herring worm (*Anisakis simplex*) has four larval stages plus the adult stage. Final hosts for this worm are dolphins, porpoises, and sperm whales. The codworm or sealworm (*Pseudoterranova decipiens*) also has four larval stages plus an adult stage. For this worm the final hosts are gray seals, harbor seals, sea lions, and walruses. The life cycle of the tapeworm begins with eggs passed into the water in feces. The coracidium hatches in the water from the egg. The coracidium is ingested by a crustacean and develops into a proceroid. This stage gets into the fish when the fish ingest the crustacean. The plerocercoid stage, which develops in the fish, infects humans and birds when they eat the fish.

Although humans are not hosts for any of the stages in the life cycle of nematodes, human infections occur when fish are consumed that contain nematodes at a stage where they can cause human illness. Man is a suitable host for tapeworms and they can mature and cause disease in humans. There are several microsporidian and myxosporidian parasites that infect fish, in particular *Henneguya saliminicola* in Pacific salmon and *Kudoa* spp. in Pacific hake and arrowtooth flounder. These protozoan parasites are of no public health concern as they do not infect humans. However, they do cause quality problems in texture and appearance of the fish.

While thorough cooking kills nematode and cestode worms, seafoods are often eaten raw or not completely cooked. Foods that have been implicated in human infections include the following (USDA 2001):

- *Ceviche (fish with spices marinated in lime juice).*

- *Lomi lomi (salmon with onion and tomato marinated in lemon juice).*

- *Poisson cru (fish with onion and tomato marinated in citrus juice and coconut milk).*

- *Herring roe.*

- *Sushi (raw fish with rice).*

- *Sashimi (raw fish).*

The reason for including this discussion on parasites in a freezing manual is that nematode and cestode worms of public health concern are sensitive to freezing. The susceptibility to freezing as a method to kill parasites depends on the species of parasite. As a rule, tapeworms are more susceptible to kill by freezing than roundworms which are more susceptible than flukes. To be sure potentially pathogenic worms are killed by freezing, the FDA has established three time-temperature regimes as follows:

- *Freezing and frozen storage at −4°F (−20°C) or below for a total time of 7 days.*

- *Freezing at −31°F (−35°C) or below until solid and storage at −31°F (−35°C) or below for 15 hours.*

- *Freezing at −31°F (−35°C) or below until solid and storage at −4°F (−20°C) or below for 24 hours.*

These time-temperature conditions may not be suitable for large fish (thicker than 6 inches).

It has been reported that all stages of the nematode *Trichinella spiralis* will be killed when frozen to a temperature of 0°F or lower. However, Alaska Cooperative Extension Service recommends that cooking be used to destroy this parasite rather than freezing. The concern is that the exposure of this parasite to cold during its lifetime has made it more resistant to destruction by freezing.

Crystal Formation, Crystal Growth, and Recrystallization

As the temperature is lowered below 32°F (the freezing point of pure water), ice will crystallize out of the tissue fluid. As ice forms, the solutes are concentrated. With concentration of enzymes and other molecules, the loss of quality due to enzymic and chemical reactions is most rapid in this temperature zone. This is called the critical zone (considered to be 28 to 23°F) and it is important to get through this zone as quickly as possible. As ice is formed, the liquid phase is concentrated. The viscosity of the liquid phase is increased by both increasing concentration and decreasing temperature. Continued ice formation will result in a state where the liquid has such restricted mobility that it is no longer possible for ice to crystallize. The phase from which no more ice can crystallize has the characteristics of a glass.

Ice is formed by a two-step process. The steps are nucleation and propagation. There must be a seed upon which the solid phase grows. The seed is a cluster of molecules with enough size to support growth. Nucleation is the formation of these seeds. The limitations on freezing are

- *Limitations on nucleation.*

- *Limitations on rate of heat transfer.*

- T_g.

Below T_g no more ice can form. But if the temperature rises above T_g, more ice can be added (Reid 1993).

The size, shape, and place (intracellular or extracellular) of ice crystals in fish muscle depend on rate of freezing and stage of rigor. Under commercial freezing conditions, large intracellular columnar crystals will be formed. Ice crystals in frozen foods can be visualized using an x-ray microcomputed tomography system (Mousavi et al. 2005).

In plants, the cell walls or cell membranes are strong and provide a physical barrier to ice propagation (propagation through the barrier is prevented). Thus in plants most ice is formed extracellular, and water moves out of the cell by osmosis to be added to ice crystals. In animals there is not a strong cell wall. The cell membrane is much less effective in preventing ice crystal formation, so internal freezing occurs more readily than in plant cells. Therefore, the rate of freezing fish is not so important as it is with plants.

When fish muscle is thawed shortly after freezing and the storage conditions have been good (low temperature and minimum temperature fluctuations), the muscle fibers will resorb almost all of the meltwater. If the storage conditions are not so good, the water will be only partly resorbed.

Effect on Flesh Proteins

The commercial value of seafood products depends mainly on the solubility of the protein. This is affected by many factors including pH, salt concentration, formation of free fatty acids, and breakdown of trimethyl amine oxide to form formaldehyde. This loss of solubility is caused by protein denaturation, and fish muscle protein is especially sensitive to this (Reid et al. 1986).

Fish has two types of muscle. The dark muscle occurs in greater amounts in strong swimmers such as tuna and mackerel, and less in more sluggish fish such as cod and flatfish. Light muscle makes up the major part in all fish species. The sarcoplasmic proteins are water-soluble and include the enzymes and oxygen carriers. The myofibrillar proteins are soluble in up to 0.3 M salt and include myosin, actin, tropomyosin, and troponin. The stroma proteins are not soluble except in strong salt solutions or if cooked, and include collagen and elastin.

Freezing and frozen storage can lead to considerable decrease in protein solubility and result in poorer texture (toughness) and increased dryness. This lower quality has been described by sensory panels as tough, chewy, rubbery, stringy, crumbly, and fibrous. Lipid oxidation compounds are involved in these changes as they interact with proteins and lead to protein denaturation. Following is a listing of changes due to freeze denaturation.

- *Unfolding of native protein molecules.*

- *Interactions between reactive groups of different protein molecules or other muscle components.*

- *Cross-linking induced by formaldehyde.*

- *Aggregation of proteins.*

- *Viscosity changes.*

- *Loss of protein solubility.*

- *Loss of water retention (lower water binding capacity).*

- *Loss of gel-forming ability.*

- *Lower lipid-emulsifying capacity.*

- *ATPase activity is lost.*

Myofibrillar protein denaturation and aggregation results in changes in texture and muscle water-holding capacity. The aggregation is due to the formation of secondary (new) bonds. The types of secondary bonds formed are

- *Electrostatic (ionic) bonds.*

- *Hydrogen bonds.*

- *Hydrophobic interactions.*

- *Disulfide bonds or bridges.*

- *Non-disulfide covalent bonds.*

The optimum temperature for this denaturation seems to be in the temperature range of 30 to 28°F, so when freezing fish it is important to get through this zone as quickly as possible.

Several theories on why protein is denatured during freezing have been advanced. The three most compelling are

- *Salt concentration and denaturing due to freezing out of water.*

- *Water molecules surrounding the protein molecules change the conformation of the protein.*

- *Free fatty acids formed affect myofibrillar protein stability.*

A report on the thawing and refreezing of cod fillet blocks (Hurling and McArthur 1996) showed there was a faster loss of myofibrillar protein solubility for twice-frozen blocks than for once-frozen. However, sensory analysis of cooked fillets after 9 months' storage showed no large loss in quality.

In frozen cod and pollock, there is rapid protein denaturation due to less-stable muscle protein and the formation of formaldehyde, which reacts with the protein. The demethylase enzyme that catalyzes the breakdown of TMAO (trimethylamine N-oxide) to dimethylamine and formaldehyde occurs in several species of the order Gadiformes. The concentration of this enzyme is low in the flesh but high in the blood and kidney. In some fishery products (dried and preheated frozen), the nonenzymatic formation of formaldehyde occurs. It has been reported that in Alaska pollock frozen just after spawning, the protein deteriorates more rapidly during frozen storage.

Texture Changes during Freezing

As discussed in the previous section on protein changes, the denaturation of the protein is followed by aggregation, which involves cross-links between protein molecules. When this happens, the denatured protein becomes much tougher than undenatured protein, and there will be more thaw drip. In extreme cases the cooked fish will be dry and the texture will be described as chewy, rubbery, or stringy.

These texture changes occur mainly during storage rather than during the freezing process. However, if the freezing process is very slow and the time in the critical zone is long, appreciable protein denaturation and texture change will occur during freezing.

The rates of these changes are much faster in minced fish than in fillets or gutted whole fish. For this reason, cryoprotectants are very important additives for minced fish or surimi production.

In gadoids, the breakdown of trimethylamine N-oxide to dimethylamine and formaldehyde results in toughness due to protein aggregation. This is caused by interaction of the formaldehyde with the muscle protein. The enzyme that causes the TMAO breakdown is active down to as low as –18°F (Regenstein et al. 1982).

Thaw Drip and Cook Drip Losses

Drip (occasionally referred to as purge) is the expressible fluid from fish flesh that forms during thawing or cooking. The volume increases dramatically with protein denaturation. There is a direct correlation between drip and texture. Excessive drip loss during thawing and/or cooking will result in dry, stringy texture.

Thaw drip and cook drip losses are reported as a percentage as follows:

% thaw drip loss =
$\frac{(weight\ before\ freezing - weight\ after\ thawing)}{(weight\ before\ freezing)} \times 100$

% cook yield = $\frac{(weight\ after\ cooking)}{(weight\ before\ cooking)} \times 100$

Juiciness is determined by the amount of moisture released during consumption.

High drip loses are undesirable because

- *They are visually unattractive.*
- *Soluble nutrients are lost.*
- *Flavor compounds are lost.*
- *Weight loss is an economic loss.*
- *Results in dry, stringy texture.*

The factors that affect the amount of drip include

- *Formation and size of ice crystals in the tissue.*
- *Location of ice crystals.*
- *Rate of thawing.*
- *Internal pressure during freezing.*
- *Irreversibility of water removal from cells (extent of water resorption).*
- *Physiological status of tissue prior to freezing.*
- *Intrinsic water binding prior to freezing.*

Changes in number and size of ice crystals during frozen storage are similar in plant and animal systems. They occur by

- *Ostwald ripening (recrystallization).*
- *Accretion or sintering.*

In Ostwald ripening, the number of ice crystals is reduced and the size is increased. It occurs at both constant and fluctuating temperature, but is especially important when there is a fluctuating temperature. This process correlates with drip loss. Accretion or sintering is when contacting ice crystals join together giving an increase in crystal size. The process is most marked with very small crystals, but happens with any size crystals.

Sutton et al. (1994) developed and tested a mathematical model of recrystallization kinetics that can be used to predict ice crystal size distri-bution in frozen foods, as affected by time and temperature. The mechanisms of migrating recrystallization (Ostwald ripening) and accretion were examined.

Thaw drip values can often be 3 to 5%. In beef, thaw drip values can be as high as 12%, and they have been found to be over 18% in fish (Kramer and Peters 1979).

Dehydration and Moisture Migration

Dehydration of the product in the cold store is known as freezer burn. If the product is not protected by suitable packaging or by a layer of ice (glazing), or if this layer of ice has sublimed away, the surface of the product will become porous and spongy. If severe, this dehydration will affect the entire surface and can go deep into the product. The dried protein will be extensively and irreversibly denatured.

With air-blast freezing, a weight loss of 1-2% can be expected. If the freezing is inefficient, there can be significant freezer burn.

After freezing to –4°F, fish muscle has water as

- *Ice.*
- *Unfrozen free (freezable).*
- *Unfrozen not free (unfreezable).*

The amounts of the different types of water have been calculated at three different temperatures: –23°F, 5°F, and –22°F (Storey 1980) (Table 1-2).

In slow freezing, the fluid in the extracellular space freezes first and the solutes are concentrated. Water is then drawn out osmotically from the still unfrozen cell. A photomicrograph of slowly frozen post-rigor cod (Love 1966, 1988b) (Figure 1-4) shows dehydrated cells around large ice crystals caused by water diffusing out to help form the large crystals. Temperature gradients in the product due to temperature fluctuations may also result in moisture migration. Temperature gradients reverse in cycles, and this results in moisture moving to the surface, but not back to the denser part of the product. Consequently, moisture collects at the surface of the product, on internal surface of the package, and in voids in the package.

Moisture migration can be prevented by using cryoprotectants.

Table 2-2. Apparent T_g' values for muscle and muscle protein fractions

Sample type	T_g' (°C) Observed[a]	Repl.[b]	Published value
Mackerel			
Whole muscle	−13.3±0.5	3	none
Soluble protein fraction without salts[c]	−8.3±0.6	2	none
Insoluble protein fraction without salts[c]	−7.5±0.4	3	none
Cod			
Whole muscle	−11.7±0.6	3	−77 (Nesvadba 1993)
			−15[d] (Simatos & Blond 1993)
			−15[d] (Levine & Slade 1989a)
Insoluble protein fraction without salts[c]	−6.3±0.1	2	none
Beef muscle	−12.0±0.3	3	−40[e] (Simatos et al. 1975)
			−60 (Rasmussen 1969)
			>−5[d] (Levine & Slade 1989a)

Source: Drake and Fennema 1999

[a]DSC procedure involved annealing samples for 1 hr at −15°C. Since the annealing procedure was not optimized for each sample type, the T_g values are probably slightly below the corresponding T_g' values.
[b]Number of replicates. Variance from mean is standard deviation for triplicates and range for duplicates.
[c]MW cut-off: 6000-8000.
[d]Estimates.
[e]Onset of ice melting as determined by differential thermal analysis.

Glassy Phase and Glassy Transition

As the temperature of a seafood product reaches just below the freezing point, ice crystals will form. The material between the ice crystals from which the ice crystals are forming is a liquid. As this liquid gets more concentrated, it becomes thicker or more viscous. This is due to lower temperature and to increased concentration. Flow becomes more difficult, until there is almost no flow. At this point it is called a glass (Nesvadba 1993, Simatos and Blond 1993). A glass is a liquid with no mobilities. Movement of molecules is much more restricted than a liquid, and glass is essentially a solid. However, it is not crystalline and does not have a regular structure. The glass is nonequilibrium, metastable, amorphous, and disoriented. The viscosity is extremely high (10^{10} to 10^{14} Pas) and there is very registered mobility.

Glass transition is associated with this unfrozen phase (UFP). The UFP goes from a viscoelastic liquid to a brittle, amorphous, solid glass. The glass transition is a glass-liquid transition, where glass is in equilibrium with supercooled melt (Le Meste et al. 2002). In food, these changes are especially important in freezing and freezedrying.

There are several reasons why glass transition is important in food preservation (Goff 1997, Levine and Slade 1989), the most important being

- *Quality loss is greatly reduced when UFP is in the glassy state.*

- *Use of the Williams-Landel-Ferry model to describe reaction rates depends on knowledge of ΔT (between the storage temperature T_s and the glass transition temperature T_g).*

- *Knowledge of environment factors on glass transition make it possible to enhance shelf life in production of frozen food.*

- *T_g can be used to assess the stability of food to deterioration resulting from diffusion-limited events both physical (ice crystal growth, solute crystallization) and chemical (enzyme reactions, protein denaturation, vitamin loss).*

Glass occurs over a defined temperature range. T_g is the temperature at which glass forms or where the glassy state is changed to the rubbery state. It is in reality a temperature range. Below T_g is the rigid glassy state where the internal viscosity is high. Molecular mobility and diffusion are very low, and chemical reactions are virtually nonexistent. Above T_g the unfrozen water is in a softer, viscoelastic rubbery state with lower viscosity and higher mobility. The highest storage stability is below T_g as reaction rates will be very slow. Below T_g no more ice will crystallize out.

The maximally freeze-concentrated T_g is denoted T_g'. This is T_g at its highest attainable value. This is where ice formation ceases within the time scale of normal measurement. Knowing T_g' is important in understanding storage stability since below this, changes will be extremely slow. This temperature is called the maximally freeze-concentrated glass. Table 2-2 gives some T_g' values (Brake and Fennema 1999).

The determination of T_g can be done in the following ways.

- *Calorimetry (Bell and Touma 1996, Simatos et al. 1975).*

- *Mechanical relaxation (Reid undated).*

- *Pulse nuclear magnetic resonance (Ruan et al. 1999).*

To obtain the highest storage stability, the storage temperature should be kept below T_g as this is where reaction rates are very slow (Reid 1990). T_g can be altered by adding certain polymers that have a high T_g as this will raise T_g for the mixture.

Watanabe et al. (1996) reported on compressive fracture stress and the glass transition temperature in frozen bonito and in a dried bonito product called katsuo-bushi.

Internal Pressure Effects

The theory has been advanced that for larger fish, the outer layer freezes first and forms a hard shell. This shell would resist expansion of the inner tissue as it froze.

It has been shown in beef that pressure measurements show sudden pressure drops during freezing, likely due to ruptures in the muscle. But in practice, there are few visible ruptures, and it is concluded that in most cases the tissue remains plastic enough to permit expansion during the freezing process (Jul 1984).

However, calculations have shown that considerable pressure can result when large fish are frozen quickly (Sternin 1991). The experience of a Kenai Peninsula, Alaska, salmon processor using liquid nitrogen to freeze chinook salmon was serious loss of quality. The fish were curled up, cracked, and full of holes (Ocean Leader 1981). This was an example of freezing too fast. High internal pressure may account for the report that pink salmon showed considerably more bruising after being frozen than before freezing.

Freeze-Cracking

Rapid freezing can give very high quality seafood. The quality of the products can be close to very fresh quality. However, with cryogenic freezing using an extremely low-temperature liquid, you can get cracking or shattering. Freeze-cracking is due to one or more of the following.

- *Volume expansion—water expands 9%; food also expands but not that much.*

- *Relieving of internal stress—nonuniform contraction or expansion.*

- *Crust or shell formation in the outer layer at the product surface.*

Seafood is most commonly frozen using air-blast, contact plate, or brine immersion. Air-blast uses cold air blown over the product; with contact freezing, the product is put on or between cold metal plates; and with immersion freezing, the product is put into cold brine. The fourth method is cryogenic freezing, where the product is exposed to a cryogenic medium at –76°F or lower. This is accomplished by immersion in a liquid, direct spray onto the food, or blowing a vapor over the food. Liquid nitrogen (boiling point is –320°F) and solid or liquid carbon dioxide are usually used. When using carbon dioxide, operators should not be in levels of 0.5% or higher for a long exposure time.

The **advantages** of cryogenic freezing are

- *Fast freezing means a short time in the critical freezing zone (shellfish meats and small fillets are frozen in a few minutes).*

- *High-quality products can be obtained.*

- *Low equipment costs.*

- *Small space needs.*

- *Product weight loss is low.*

  The **disadvantages** of cryogenic freezing are

- *Transport of liquid nitrogen is costly.*

- *Operating cost will be low if the cryogenic medium is readily available, but could be high if it is not (the operating cost could be as much as three times higher than air-blast or contact plate freezing).*

- *Processing area must be well ventilated.*

- *Operator could be asphyxiated if the oxygen level is reduced below 10%.*

- *Freeze-cracking is more likely to occur than with air-blast or contact plate freezing.*

Statistical analyses of 10 physical properties for 22 different materials showed that several of these properties correlated significantly with freeze-cracking (Kim and Hung 1994). A set of equations was developed to predict cracking incidence during cryogenic freezing. Susceptibility of several food commodities to freeze-cracking was determined.

Modeling to predict freeze-cracking susceptibility depends on knowledge of heat transfer and stress analysis (Hung and Kim 1996). This will allow identification of a freezing rate that will not cause freeze-cracking. Another approach to avoid freeze-cracking is to use thermal equilibrium freezing, during which the internal temperature of the product is reduced to between 30°F and 28°F in a 5°F cold room subsequent to freezing. This procedure was used to prevent freeze-cracking during brine freezing of tuna (Ogawa 1988).

Gaping

Gaping occurs when the connective tissue in fish muscle (myocommata) fails to hold the muscle segments or blocks (myotomes) together. The surface of a fillet looks split or cracked. If severe, a skinned fillet will fall apart. The severity depends a lot on species and nutritional condition of the fish. Cod and salmon are very susceptible to gaping. Very little gaping is found in flatfish. Well-fed fish are more apt to gape.

The main causes of gaping are high temperature or bad handling during the process of rigor. Gaping can be caused by bending or straightening a curved fish while the fish is in rigor.

Freezing causes some gaping and refreezing causes more gaping. Love (1988b) reported that fish frozen very slowly (over a 3-day period) gape more than those frozen very fast (in 8 minutes) or moderately fast (in 3 hours). The best results are with fish frozen pre-rigor.

Effects of Rigor Stage

The process of rigor is a stiffening of the body of an animal after death due to contraction of the skeletal muscles. It can be described as follows:

limp, pliant muscle, pre-rigor $\xrightarrow{\text{1–12 hours}}$ contracted, rigid muscle, in-rigor $\xrightarrow{\text{24 hours or longer}}$ muscle relaxed and soft, post-rigor

The times given for the progression through the stages of rigor are very rough approximations. Many factors affect these times including size of fish, species, nutritional condition, temperature, and handling. The enzymatic reactions associated with rigor mortis involve nucleotide changes, which proceed in the following sequence:

$ATP \longrightarrow ADP \longrightarrow AMP \longrightarrow IMP \longrightarrow Inosine \longrightarrow Hypoxanthine$

Formation of hypoxanthine is associated with bitter flavor.

More stress during capture or delay in chilling will give a faster onset of rigor. Also, rigor will last longer if there is less exertion of the fish before death and if the fish is refrigerated soon after death. When trawled fish are exhausted by struggle in the net or on deck, they will go into rigor quickly after death, sometimes within minutes.

Strength of rigor is dependent on temperature. Morrison (1993) reported that the maximum strength of rigor in cod occurs at 63°F.

There are reports that freezing pre-rigor gives lower quality than freezing post-rigor. Others say that freezing pre-rigor gives very high quality (Martinsdóttir and Magnússon 2001). Pacific halibut frozen pre-rigor has been shown to have better flavor and odor, lower free drip, and longer frozen storage life compared to halibut frozen after storage in ice or refrigerated seawater (RSW) (Tomlinson et al. 1973). It is generally accepted that there is less deterioration during frozen storage with fish frozen pre-rigor when the fish is on the bone. However, when freezing fillets, stage of rigor is more important than when freezing the whole fish (due to the absence of the frame for support to prevent shrinkage). If fillets taken pre-rigor enter rigor before freezing, there can be considerable shrinkage (up to 40%) since there is no skeleton to anchor and pull against. It has been reported that light muscle can shrink up to 15% and dark muscle up to 52% in fish filleted pre-rigor. Pre-rigor fillets should not be held in freshwater prior to freezing, as this results in increased shrinkage. Another disadvantage of freezing pre-rigor or in-rigor is the opening of the belly cavity during freezing. This makes the fish more difficult to handle and to store.

Bad handling or freezing of fish or fillets during the time they are in-rigor will result in toughening of the muscle, darker color, and loss in yield. Loss in yield can be minimized by a 10- to 20-second dip in 3 to 6% brine containing sodium tripolyphosphate or sodium hexametaphosphate. This will reduce thaw drip.

Fish in rigor in a bent position should not be straightened when put in the freezer. Straightening a curved or bent fish before rigor is complete and the muscle is relaxed will result in severe gaping. If curved during rigor, the flesh on the outside of the curve will be subject to the most strain. Fish should be frozen pre-rigor or allowed to complete rigor before being frozen.

Work with cod showed that fillets frozen after going into rigor and subsequently thawed were discolored, had ragged edges, and the texture was poor. It was concluded that cod fillets should be frozen pre-rigor or post-rigor but never in-rigor. The important point is that when fish or fish fillets are frozen pre-rigor or in-rigor, the process of rigor must be allowed to proceed to completion prior to thawing or cutting into fish sticks. The process of rigor must be completed slowly during frozen storage so that rigor does not occur during thawing (thaw rigor). If thawed or cooked before rigor is complete, the muscle will contract, with a loss of cellular fluid and lots of thaw drip and cook drip. Thaw rigor results in tough muscle tissue and a lot of gaping. To avoid lots of thaw drip, rubbery texture, and serious shrinkage (with fillets), fish or fillets frozen solid before the onset of rigor mortis must be thawed slowly at a low temperature. The energy source for muscle contraction during thawing is ATP. You can get formation of ATP from glycogen (glycolysis) during thawing. Thaw rigor can be avoided by ending ATP synthesis and degrada-

tion at subfreezing temperatures while the muscle is still frozen.

Another approach to avoid thaw rigor is by conditioning. Conditioning is when the fish is held at a relatively high temperature in the freezing process to allow completion of rigor while the muscle is still frozen.

Cryoprotectants

Several undesirable changes take place during freezing, frozen storage, and thawing. In particular, freeze denaturation of muscle proteins gives

- *Changes in conformation of myofibrillar proteins.*

- *Formation of insoluble protein complexes.*

- *Reduction of enzymatic activity of myosin and sarcoplasmic proteins.*

Fish muscle proteins are less stable than those of beef, pork, or poultry. The most susceptible to freeze damage are the myofibrillar proteins (myosin). The least susceptible are the sarcoplasmic proteins (globulin, myogen, and myoalbumin, and the stroma proteins (collagen and elastin).

Native protein conformation can be stabilized by the addition of several types of cryoprotectants, including sugars, polyalcohols, carboxylic acids, amino acids, and polyphosphates. One of the most common cryoprotectant combinations is sucrose, sorbitol, and a polyphosphate. The sweetness of this mixture is unacceptable to some consumers, so there has been a lot of work on other cryoprotectants (Jittinandana et al. 2005). An excellent review on this subject has been published by Park (1994).

The action of cryoprotectants improves water holding to prevent water migration and increases surface tension. The mechanism of stabilizing proteins is through cryoprotection and cryostabilization. Cryoprotection is where low molecular weight compounds favor thermodynamically the maintenance of proteins in their native state. Cryostabilization involves use of high molecular weight polymers to raise the glass transition temperature.

Sugars and other carbohydrates are widely used as anti-denaturants. Sucrose is a common cryoprotectant used in surimi production. The most effective sugars and other carbohydrates are sucrose, lactose, galactose, glucose, fructose, and maltose. Also effective are mannose, trehalose, fractose-6-phosphate, and polydextrose. Less effective are ribose, xylose, raffinose, and glucose-phosphate. Starch has little or no effect as a cryoprotectant.

Several polyols or polyalcohols and related compounds are used to stop or retard protein denaturation. The most commonly used is sorbitol, but xylitol and glycerol are also effective. Propylene glycol is moderately effective but mannitol, erythritol, glyoxal, and dihydroacetone are not good cryoprotectants.

Carboxylic acids that are highly denaturant are malonic, methylmalonic, glutaric, glyceric, maleic, L-malic, tartaric, gluconic, citric, and 2-amino-butyric. Moderately effective are D,L-malic, and odipic. Fumaric, succinic, and oxalic afford little or no protection.

Studies on protection of protein from freeze damage show that the amino acids aspartic acid, glutamic acid, cysteine, and glutathione were highly protective. Lysine, histidine, serine, alanine, and hydroxyproline gave moderate protection while glycine, leucine, isoleucine, phenylalanine, tryptophane, threonine, glutamine, asparagine, and ornithine gave low or no protection.

Some species of rockfish and flatfish show high drip loss at certain seasons, especially from cut surfaces such as fillets or portions. Polyphosphates alone or in combination with a sugar and/or a polyol are of help in improving water retention when thawed. A dip in a 10% solution of polyphosphate will result in less drip loss and less gaping. A solution below 5% is not very effective.

Rudolph and Crowe (1985) showed that lobster muscle has two natural cryoprotectants—proline and trehalose. Both are more effective than glycerol and dimethyl sulfoxide in preserving membrane structure and function during freezing and thawing.

Other compounds that have been shown to provide protection against damage during freezing and thawing alone or in mixtures are some proteins (for example, insulin) and ethylene diamine, ethylene diamine tetra-acetic acid, creatine, pyrophosphoric acid, monosodium glutamate, maltodextrins, modified starch, and several antioxidants (Hall and Ahmad 1997).

Auh et al. (1999) showed that an 8% (w/v) solution of oligosaccharide mixture was effective in cryoprotection of an actomysin solution extracted from Alaskan pollock. They suggested the mixture had good potential as a non-sweet cryoprotectant of fish protein.

Packaging

Seafood can be packaged whole, eviscerated, as IQF fillets, in shatterpacks, or as blocks. Protection by packaging consists of an outer layer (totes, cardboard cartons, or waxed cartons) and an inner layer (plastic liners, bags, or wraps). For fatty fish, polyethylene, waxed paper, cellophane, or combinations of these are not good as an inner layer as they are not gas resistant or do not provide a tight fit. With a loose fit, there will be excessive dehydration and oxidation. Therefore, a vacuum pack will be needed and a purge with nitrogen gas can be used to remove oxygen.

The problems that good packaging will protect against are weight loss, texture changes, flavor losses, nutritional losses, off odors, contamination with bacteria or other adulterants, and physical damage. The most suitable packaging material will have a low rate of transmission of water vapor (to avoid desiccation) and a low permeability to oxygen (to prevent fat and pigment oxidation). The material should be strong, tight fitting to prevent loss of moisture from the product inside the package, non-absorbing of oil or water, relatively inexpensive, easy to apply, and easy to label. Preventing desiccation is important as product weight loss could result in the net weight being low. The water vapor transfer should be no greater than 0.4 g per m^2 per 35 hours at $-4°F$ and 75% relative humidity (Sikorski and Kolakowska 1990).

Whole and eviscerated fish are usually protected by glazing (one or two dips in a glaze tank). Fillets frozen IQF (individually quick frozen) can be protected against dehydration using a dip or spray. Fillets are often placed in a shatter pack where interweaving of waxed paper or plastic film between layers allows individual fillets to be detached without thawing. Fillets are frozen packaged or unpackaged, usually in an air-blast or a plate freezer. When freezing unpackaged fillets in an air-blast freezer, they should be removed as soon as they are completely frozen, to minimize dehydration and discoloration.

Small fish such as herring or mackerel can be frozen in a block. Fillets can also be frozen in blocks. A wet block has water added; a dry block has no added water. Blocks are best frozen in water or glazed, for three reasons:

- *The coating of ice protects against physical damage.*

- *The block will be stronger.*

- *Keeping oxygen away retards rancidity and color changes.*

Glaze is a layer of ice coating the product and is reported as a percentage of the product weight. There are lots of dips or sprays that can be used to give protection in the cold store, but ice is the only one of commercial importance. A glaze is commonly used for whole and for headed and gutted fish. The glaze solution temperature should not be above 40°F and should be kept as close to 32°F as possible. Ice should be skimmed off to prevent a rough glaze. If the fish is very cold, a single dip of 45 seconds is best. For previously dipped fish, one dip of 15 seconds is sufficient. After glazing, it has been reported that fast re-freezing is important to reduce thaw drip in salmon (Bilinski et al. 1977) and in cod and trout (Nilsson and Ekstrand 1994).

For a better glaze, additives can be added to make the ice coat less brittle and reduce evaporation. These glaze additives improve the following characteristics:

- *Water holding capacity.*

- *Elasticity.*

- *Adhesiveness.*

- *Clarity.*

Additives that have been used to strengthen the glaze include salt, corn syrup solids, sugar (sucrose or glucose), starch, carboxymethyl cellulose, sodium alginate, and monosodium glutamate. Chemical properties (antioxidant potential) have resulted in the use of several compounds in glaze water including sodium erythorbate, ascorbic acid, and butylated hydroxyanisole (BHA) or butylated hydroxytoluene (BHT). It is thought that antioxidants incorporated in glaze water are not a significant help.

Edible coatings including chitosan, egg albumin, soy protein concentrate, pink salmon protein powder, and arrowtooth flounder powder have been shown to provide protection during frozen storage (Sathivel 2005). Using these edible coatings on skinless pink salmon fillets, Sathivel found improvements in thaw yield and cook yield, reduced lipid oxidation, and less moisture loss.

Another value of packaging to lower water vapor loss is not related to product quality changes.

It is the reduction of frost forming on the evaporator plates, resulting in a less-efficient refrigeration unit.

RECOMMENDED FREEZING RATES AND CORE TEMPERATURES

Freezing Rates

Fish starts to freeze at about 29°F. In general, the faster the freezing rate the better. Keeping the amount of time in the critical freezing zone short is important in minimizing chemical and enzymatic reactions that lower quality (see Figure 2-1). You should be able to freeze fish that are at 41°F down to −20°F in 4 hours or less. Jul (1984) reported there is no flavor loss in cod frozen at a rate between 0.15 cm per hour and 5 cm per hour. However, the recommended rate is not less than 0.5 cm per hour. When fish are frozen pre-rigor, the freezing rate is of almost no importance except in the case where the freezing is very slow. Nearly all the water is inside the cells, and the crystals are formed inside. If frozen in-rigor or post-rigor, some water is outside the cell, and crystals form on both sides of the cell wall. With a slow rate of freezing, crystals form in the extracellular fluid, concentrating the salts and resulting in water moving out of the cell and forming large extracellular crystals. Tomás and Añón (1990) reported that fast or slow freezing rates have no effect on the rate of lipid oxidation. Table 2-3 shows freezing times for a number of different fish products.

Slow freezing results in a small number of large ice crystals, most of which are extracellular. When freezing takes a very long time (about 20 hours), very large crystals up to 10 mm long can form. There is little doubt that a very slow freezing rate (for example, in still air at 14 to 20°F) will give inferior quality. There will be texture changes, accelerated rancidity, and higher weight loss from drip due to cell wall rupture. Freezing times longer than 6 hours are not advocated. But except for very long freezing times, the rate has almost no effect on quality of the product.

Fast freezing results in large numbers of small ice crystals that are both intracellular and extracellular. Also, the time in the critical zone is shorter. The result is higher quality with lower drip loss and good texture is maintained. Fast freezing is achieved using low operating temperatures and/or

Figure 2-1. Critical freezing time.

(Licciardello 1990)

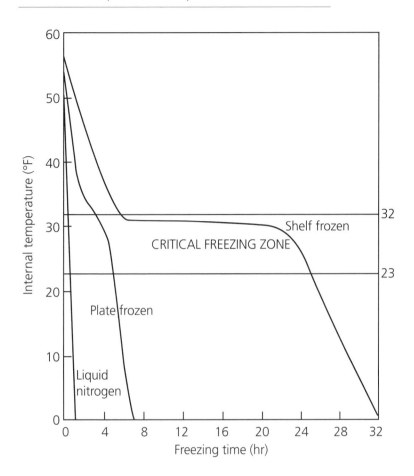

high heat-exchange rates. The driving force that determines the ice nucleation rate is supercooling. Supercooling is the difference between the actual temperature of the sample and the expected solid-liquid equilibrium temperature. For each degree of supercooling, the ice nucleation rate is increased tenfold. In the laboratory where fish were frozen almost instantly, the ice crystals formed were too small to be seen using a microscope. It should be pointed out that damage can occur from extremely short freezing times (see discussion on freeze-cracking).

Nicholson (1973) listed the factors that affect the freezing rate as

- *Type of freezer.*

- *Operating temperature (difference between product and coolant).*

- *With air-blast, the air speed.*

- *With contact plate, the product contact area.*

- *Initial product temperature.*

- *Product thickness.*

- *Product shape.*

- *Density.*

- *Species.*

- *Method of packaging and fit.*

The temperature at which fish muscle starts to freeze depends on the solutes in the tissue fluids. Some examples are

- *Cod—30.4°F.*

- *Halibut—28°F.*

- *Herring—34.5°F.*

Table 2-3. Freezing times for fish products

The following are observed times for a number of fish products; although the freezing time of a new product in a particular freezer should always be measured in the recommended manner, these typical freezing times will give designers and operators some idea of what to expect in practice.

Product	Freezing method	Product initial temp °C	Operating temperature °C	Freezing time hr	min
Whole cod block 100 mm thick	Vertical plate	5	−40	3	20
Whole round fish 125 mm thick eg cod, salmon, frozen singly	Air blast 5 m/s	5	−35	5	00
Whole herring block 100 mm thick	Vertical plate	5	−35	3	20
Whole herring 50 mm thick on metal tray	Air blast 4 m/s	5	−35	1	40
Cod fillets laminated block 57 m thick in waxed carton	Horizontal plate	6	−40	1	20
Haddock fillets 50 mm thick on metal tray	Air blast 4 m/s	5	−35	2	05
Haddock fillets laminated block 37 mm thick in waxed carton	Horizontal plate	5	−40	1	02
Kippers in pairs interleaved pack 57 mm thick in cardboard carton	Horizontal plate	5	−40	2	15
Whole lobster 500 g	Horizontal plate	8	−40	3	00
Whole lobster 500 g	Liquid nitrogen spray	8	−80	0	12
Scampi meats 18 mm thick	Air blast 3 m/s	5	−35	0	26
Shrimp meats	Liquid nitrogen spray	6	−80	0	5

Source: Nicholson 1973

An important rule is to never place unfrozen or partially frozen fish into the cold store. The freezing rate unfrozen or the interior of partly frozen product will be too slow for high quality. Also, the temperature of the cold store will go up.

FMC Foodtech of Sweden manufactures two Frigoscandia impingement freezers (Wray 2005). Table 2-4 gives the capacity and performance of one of these freezers. They are advertised to freeze five to six times faster than a conventional spiral freezer and to be at half the cost of cryogenic freezing. This very fast freezing method uses thousands of high velocity jets of air directed at both sides of the product, to strip away the insulating boundary layer of air. It provides very fast freezing of thin products like shrimp, squid rings, scallops, and fish fillets.

With the current interest in improved quality of seafood products, high-pressure shift-freezing (PSF) is becoming of interest as a method of freezing seafood (Préstamo et al. 2005). It permits uniform cooling of the whole volume of the sample. This is accomplished by a rapid release of pressure from a supercooled unfrozen product. A schematic description of an experimental system for pressure-shift freezing is shown in Figure 2-2. PSF can give improved quality compared to air-blast, spray or liquid immersion, and plate or contact freezing. It preserves microstructure better and creates smaller, more uniform ice crystals. There remains some question as to whether it is suitable for muscle tissue. Chevalier et al. (2000) compared PSF and air-blast freezing of Norway lobster. Their scanning electron micrographs of the frozen tissue showed PSF yielded smaller ice crystals.

Frozen storage fluctuations have more influence on quality deterioration than does freezing rate (Aurell et al. 1976). However, freezing rate can be very important if the frozen storage conditions are less than ideal.

End Core Temperatures

The end core temperature should be −5°F or lower and should always be lower than the temperature of storage. End core temperatures in the range of

Table 2-4. Capacity and performance of Frigoscandia's Advantec Impingement Freezer

Product	Infeed temperature	Freezing time in minutes	Belt load kg/m (lb/ft)	Cap./module kg/hr (lb/hr)	Weight loss (%)
Cooked shrimp 1 layer, 13 mm (½ inch) thick	+25°C (+77°F)	3.1	4.7 (3.15)	410 (904)	0.7
Cooked shrimp 2 layers, 13 mm (½ inch) thick	+25°C (+77°F)	3.9	9.4 (6.3)	655 (1444)	0.7
Raw redfish fillets 15 mm (½ inch) thick	+5°C (+41°F)	5.0	5.7 (3.8)	320 (705)	0.9
Raw redfish fillets 23 mm (1 inch) thick	+5°C (+41°F)	7.9	7.5 (5.0)	280 (617)	0.9
Raw cod portion 23 mm (1 inch) thick	+5°C (+41°F)	8.0	9.5 (6.4)	320 (705)	0.9

Source: Wray 2005

Figure 2-2. Schematic description of an experimental system for pressure-shift freezing.

(from Zhu et al. 2004)

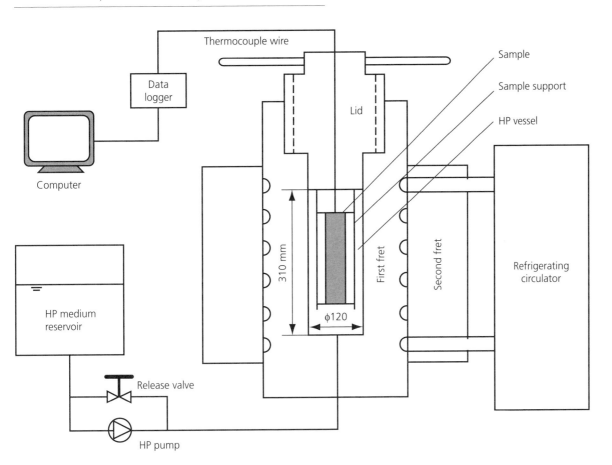

15°F down to –5°F are not recommended. There will be more damage and greater release of enzymes than at lower end core temperatures.

There is a theory that end temperature should not be too cold. It is argued that at very cold temperatures some chemically bound water is removed. The reversibility during thawing for this water to go back to its original position in tissue would be jeopardized. Therefore the reversibility of the freezing-thawing process is not attained and quality is lost. However, for some products (an example being sushi-grade tuna), end core temperatures of –40 to –70°F and lower have been recommended.

Precooked vs. Raw

Seafood is considered to be a fragile protein food. It is generally cooked for shorter times than other muscle foods. Cooking for longer times can cause it to toughen and become dry. Cooking firms up the product, due to protein denaturation and water loss. A lower water content means a lower freezing requirement compared to raw product.

FREEZING SEAFOODS AND SEAFOOD PRODUCTS

Warmwater vs. Coldwater Species

Pseudomonad bacteria dominate in the spoilage of North Pacific fish. These spoilage bacteria are psychotropic or cold-resistant and are associated with fish found in cold or temperate water but not in tropical water fish.

Since cold and temperate water fish have cold-resistant spoilage bacteria, icing tropical fish has a greater effect in reducing spoilage than does icing of cold or temperate water fish. This means chill storage extends the shelf life more for warmwater fish, and it explains why they have longer keepability.

With coldwater fish, the onset of rigor is faster at higher temperature. With warmwater fish, the onset of rigor is faster at 32°F.

Most freezing and frozen storage studies have been done with fish of north temperate waters (Haard 1992). It appears that both tropical and Antarctic species are less vulnerable to texture deterioration when frozen than are north temperate species.

Fish and Fish Products

Lean Fish

Most lean fish or shellfish have about 1% lipid or less. Almost all of this occurs as phospholipid. Fish or shellfish with 2% or less lipid are classified as lean. Species considered lean include cod, pollock, flounder, grouper, hake, haddock, Pacific ocean perch, Pacific halibut, crab, oysters, clams, scallops, and squid.

Compared to fatty fish, lean fish have a higher water content and therefore a higher freezing requirement. The major storage problem is toughening of the muscle.

The fish may be frozen as

* Round (whole).

* Drawn (gutted or eviscerated only).

* Dressed (gutted or eviscerated and trimmed).

* Fillets.

To obtain a high-quality frozen product, ocean whitefish should be held in chill storage no longer than 2 to 3 days, and flatfish no longer than 3 to 5 days, before freezing.

Fatty Fish

Fish and other seafood can have a very wide range of fat content. It often varies up to 20% and in some cases up to 30%. Over 5% lipid is considered to be fatty. Examples of fatty fish are herring, sablefish, mackerel, eel, dogfish, chinook salmon, sockeye salmon, albacore tuna, and rainbow trout.

There is a wide variation in lipid content, depending on time of year. For example, mackerel will have 20 to 25% lipid prior to spawning and about 5% after. Herring can have more than 23% oil prior to spawning, and this can drop to below 2% after spawning. A fat content of at least 8% is desirable for herring flesh to be used as human food.

About 1% of the lipid in fish muscle occurs as phospholipid, much of which is associated with the cell membrane. Most of the rest occurs as triglyceride in fat deposits. In vegetable oils, there are usually one or two sites of unsaturation. In fish oil there can be up to six sites of unsaturation. Consequently, fatty fish can undergo rapid auto-oxidation. These oxidative changes result in color, odor, and flavor changes that lower quality. Fatty fish should be chilled as soon as they are caught. Freezing reduces free water, and the solutes are concentrated. This results in higher reaction rates as the freezing process progresses. Rancid flavors can develop during freezing and frozen storage if fatty fish are held more than 1 day in ice. Ideally, fatty fish should be frozen not more than 12-18 hours after capture.

In most species, as the fat content becomes higher, the water content goes down. This means a higher lipid content results in a lower freezing requirement.

Smoked Fish Products

Hot-Smoked

Historically, smoking of fish was a preservation method. It depended primarily on reduction of water activity by removing moisture and adding salt. To a lesser degree, the antibacterial properties of some of the chemicals in the smoke have a preservative effect. Today the main reason to smoke fish is to add flavor, and the brine and smoking procedures used are much milder.

The process of hot-smoking fish gives a fully cooked product. Since the tissue is cooked, freezing and thawing is not damaging to a hot-smoked product.

Hot-smoked fish that are intended to be frozen should be put into the freezer as soon as they are cool. Do not put hot product into the freezer, as the freezer may not be able to cope with the heat load and it is more economical to cool the smoked product first. Never use stale or old smoked fish to make a frozen product. Freeze quickly (1 to 2 hours) in a good barrier package.

While hot-smoked fish can be successfully frozen and stored frozen, the keeping quality at a given temperature will not be as good as for non-smoked. At a constant temperature of –15 to –20°F, a storage life of 4 to 6 weeks can be expected with quality remaining as good. Use of vacuum packaging can give a storage life of up to 12 months.

It is important to use fresh fish so that a good pellicle is formed (firm, glossy surface). If frozen fish is used, it must be well-frozen and well-stored. Poor freezing or poor frozen storage will result in loss of soluble protein. With poor extraction of soluble protein, a good pellicle will not be formed (dull, matte surface).

Cold-Smoked

Fish that are cold-smoked are not cooked. The desired flavor and texture are obtained by brining and low temperature smoke (not over 85 to 90°F). Freezing and frozen storage recommendations are essentially the same as for hot-smoked fish. Freezing should take no more than 1 to 2 hours. Storage in a good vapor barrier package at a constant temperature of –15 to –20°F will give a couple of months' storage life. Up to 12 months can be obtained using a vacuum pack.

Rørå and Einen (2003) studied the effects of freezing, before and/or after cold-smoking Atlantic salmon, on physiochemical characteristics of the finished product. They reported on effects of freezing the raw fish and fillets before smoking, and freezing the product following smoking. Freezing before smoking gave an increase in product due to higher water content in the final product. However, freezing before smoking also resulted in softer texture and an increased k value. Although the concentration of the pigment astaxanthin was reduced by freezing, the color intensity and lightness of the fillets increased. Gaping increased in fillets frozen before smoking. Refreezing of the finished product had little additional effect on product quality. The authors point out that they used brine injection; if other salting methods are used, the effects of freezing may differ from those found in this study.

Acidified Fish Products

Fermented foods such as yogurt and sauerkraut can be prepared by adding bacteria that produce acid (such as lactic acid bacteria), and acidified foods can be made that will get the same level of preservation by adding an acid (such as vinegar). The goal is to add enough acid so the pH is below 4.6. Below this pH, spores of Clostridium botulinum will not germinate and grow. Mild heat can be used to kill the vegetative cells. A strong heat treatment that would kill the spores is not needed.

Acids reduce the microbial tolerance to freezing, so more of them will be killed during freezing and frozen storage.

Fish Roe and Roe Products

Salmon eggs processed as the whole skein are called sujiko, the main market being Japan. The steps involved in producing sujiko are washing the skeins in saltwater, brining, draining, packing in boxes, and putting weight on the boxes during the curing step. The finished product can be stored at –5°F, but –10 to –15°F is a better storage temperature. If the eggs are frozen prior to processing, the finished product will be poorer quality. The poorer quality means less firm and the quality will depend on time and temperature the skeins are held before freezing.

Salmon eggs processed as individual eggs rather than as the intact skein are called ikura. Ikura or salmon caviar with a salinity of 3.6% starts to freeze at 18°F and is frozen at a few degrees lower (Sternin and Dore 1993). It should be stored at 4°F or lower. Japanese researchers recommend –40°F for long-term storage. Frozen storage eliminates the risk of botulism. High-quality ikura frozen in vacuum packages maintains good quality for as long as 2 years. The main disadvantage of freezing is rupturing of the eggs. In North America and Europe, there is prejudice against frozen salmon caviar. There is no such prejudice in Japan, and they use a lot of frozen product. It is often shipped frozen in plastic pails.

Herring roe is salted and frozen. The Japanese market prefers female fish frozen whole, so the roe can be processed closer to the market.

Cod roe is often smoked and this can be frozen successfully.

Freezing of fish roe, if done properly, should not seriously lower the quality for most caviar products. In particular, caviars made from small eggs like those from herring and whitefish, which have firmer outer membranes, freeze well. However, sturgeon eggs, which have a thin membrane, are easily damaged when frozen.

Mollusks

Gastropods (Univalves)

The gastropods include snails, whelks, slugs, and limpets. Most have one shell, although some have no shell or just the vestige of a shell.

The commercially most valuable of the gastropods is the abalone. The large muscular foot is cut from the shell, separated from the viscera, and washed. The foot is then sliced into steaks about one-half inch thick and the steaks are pounded with a mallet to make them more tender.

Most abalone is sold fresh, with only a small amount shipped frozen. Abalone can be frozen whole, after it is cleaned, or after it has been cut into steaks and beaten. The steaks can be frozen unbreaded or breaded. Abalone should be held on ice no longer than a day before freezing.

The salt-soluble fraction of the foot muscle of the abalone *Notohaliotis discus* is about 65% paramyosin. This is a myofibrillar protein that is found only in invertebrate muscle.

Snails are marketed frozen or canned, with or without shells. Cleaned meats are mainly a product of France.

Lamellibranchs (Bivalves)

The lamellibranchs, or bivalves, include oysters, clams, cockles, scallops, and mussels. The bivalve shell halves are held together by one or two adductor muscles. There are two parts to the adductor muscle. The part with a transparent appearance is involved in fast contraction and relaxation. This muscle cannot maintain tension for long periods. In some species, this muscle has cross-striated or oblique-striated muscle fibers. The other part of the adductor muscle has an opaque appearance. It is slow acting and can maintain tension for long periods. This is smooth muscle with no striations visible under the microscope. Bivalves are marketed in the shell, as shucked raw meats, and as shelled cooked meats.

The oyster species of most economic importance are the Pacific oyster *Crassostrea gigas* and the Atlantic oyster *Crassostrea virginica*. Oysters can be frozen IQF in the shell or as shucked meats, as meats in cans by air-blast freezing, and as meats in a waxed carton with an overwrap by contact plate freezing. Oysters freeze very well. High-quality oysters frozen and protected with a glaze or suitable packaging are almost as good as fresh. Oysters in frozen storage will undergo changes in flavor, texture, color, and drip loss, but these will not be too severe with good handling. Shucked meats store better frozen than oysters frozen in the shell or on the half shell. Freezing with the shell is not very satisfactory, as it results in adverse flavor changes. Frozen oysters darken and a slow freezing rate and/or a high storage temperature make this darkening worse. The freezing rate for shucked meats should be as fast as practical and the storage temperature should be –20°F and never higher than 0°F. If oysters are fresh at the time of freezing, they will store well for several months. If they are not fresh when frozen, the quality will be poor after 1 month of storage. Dead oysters should never be frozen. An announcement by the Florida Department of Agriculture and Consumer Services (Balthrop undated) discusses freezing as a method to improve safety of oysters to be consumed raw. Work funded by the University of Florida Sea Grant College Program showed that freezing and storing of freshly harvested oysters at extremely low temperatures results in no detectable viable bacteria remaining, thus making raw oysters safer.

Many species of clams are harvested commercially. Only a very small part of the commercial harvest is frozen. Clams are not usually frozen in the shell, but small amounts are frozen as meats. They are shucked by cutting the adductor muscle and scraping out the meat. Meats that are frozen are usually chopped or minced for use in chowders. Quality problems with the frozen product are rancidity, toughening, and thaw drip. Frozen meats go down in quality relatively fast and the texture becomes spongy or tough.

In the Atlantic, three species of scallops have commercial importance: sea scallop (*Placopecten magellanicus*), bay scallop (*Argopecten irradians*), and calico scallop (*Pecten gibbus*). In the Pacific, the commercially important species is the weathervane scallop (*Patinopecten caurinus*). Most scallops are sold as fresh meats, but small amounts are frozen. Scallop meat can be frozen using air-blast or plate freezing for very fast freezing. Evaluation of flavor, odor, and texture by a sensory panel showed no overall difference in these three freezing methods. It has been reported that at 0°F the frozen storage life is 7 to 12 months.

Five species make up most of the mussels in the North American market. Four of these are native to North America, and the green lip or green shell mussel is native to New Zealand. Three of the North American species (blue mussel, black mussel, and bay mussel) were long considered to be the same species. The fourth North American species of commercial important is the surf or California mussel. Mussels are marketed as a live product, on the half-shell, meats out, smoked meats, and meats cooked in marinades or sauces. Mussels may be canned or frozen. They are frozen in the shell, on the half-shell, IQF meats (plain or breaded), and blocks of meats. The frozen storage life is about 6 months, but if held at –20°F they can be stored up to 9 months.

Two manuals have been published on mussel processing (Warwick 1984, Price et al. 1996). The manual by Price et al. is primarily concerned with development of a HACCP (hazard analysis critical control point) plan for mussel processing. Quality control is discussed but freezing and frozen storage are not covered. The manual by Warwick was written specifically for the green or green-lipped mussel produced in New Zealand. It has sections on freezing and frozen storage. Some of the important points in these sections are

- *Processing must be planned to minimize delay in freezing.*

- *Mussels must be graded before freezing.*

- *Fast freezing is essential (core temperature reduced to below –13°F within 24 hours).*

- *Mussels should be glazed after freezing.*

- *Cold store temperature should be at –22°F or lower.*

- *Cold store temperature fluctuations should be kept to a minimum.*

- *Cold stores should not be used as freezers.*

It has been shown that seasonal variation in proximate composition of blue mussels affects the frozen storage life of the cooked product (Krzynowek and Wiggin 1979). Ablett and Gould (1986) found variation in oxidative rancidity with different muscle tissue. Digestive gland tissue gave the highest oxidative rancidity values. Use of ascorbic acid with or without a chelating agent to retard progression of oxidative rancidity was effective under some storage conditions but not others (Ablett et al. 1986).

Crustaceans

Crustaceans are in the phylum Arthropoda, class Crustacea. The commercially important species (crabs, lobsters, and shrimps) are in the order Decapoda. As with fish, which have two types of muscle (light and dark), crustaceans have two types of muscle, called tonic and phasic. Tonic muscle is made up of long sarcomeres, about 10 micrometers long. It is slow to contract and relax and can sustain a long contracture. Phasic muscle has short sarcomeres about 2-3 micrometers long. This muscle is fast acting and is easily fatigued.

Crabs and lobsters must be kept alive to processing or the meat will be of poor quality due to changes in appearance, flavor, and texture (Banks et al. 1977). Most shrimp die shortly (minutes) after capture, and large quantities are stored frozen.

Several oxidation reactions in crustaceans result in undesirable color changes. The white meat is snow white or creamy white but will become yellow with poor handing or storage. With iced shrimp, enzymes in the shrimp catalyze the oxidation of compounds (mainly phenols) into melanins. The blackened areas develop as spots or bands at the base or across the back of shell segments. The discoloration is formed on the shell, but in advanced cases it occurs on the meat. This black spot or melanosis is most serious in southern warmwater shrimp (penaeid species) and much less of a problem in coldwater shrimp (pandalid species). Another type of discoloration is the development of blue or black meat, which is termed blueing. Fading or discoloration due to enzymic oxidation of carotenoid pigment may involve the following reactions.

$$\text{beta-carotene} \longrightarrow \text{astaxanthin} \longrightarrow \text{astacene}$$
$$\text{(red)} \qquad\qquad \text{(pink)} \qquad\qquad \text{(orange-yellow)}$$

The frozen storage life for crabs and lobsters is quite variable (Banks et al. 1977). The frozen shelf life is roughly

- *Blue crab and northern lobster—limited, 2-3 weeks.*

- *Dungeness crab—moderate, 3-6 months.*

- *King crab and spiny lobster—very good, up to a year or more.*

King crab is the largest of the commercial crab species and is more suitable for freezing than the other crab species. Dungeness crab is frozen as cooked whole or eviscerated crab. High-quality crab will have well-filled legs and a bright, clean shell. It will not be recently molted or a crab that has skipped a molting period.

Blue crabs are small and not good for storing frozen as whole crab. The meat is packed into cans or plastic bags, heat pasteurized, and then frozen. Studies on the texture of blue crab in frozen storage showed an increase in toughness (Morrison and Veitch 1957). This was attributed to a constant, low-grade non-enzymic respiration involving oxidation of tissue carbohydrate. It occurred at storage temperatures as low as 1.4°F.

Lobsters frozen raw will be good up to about 6 months. Cooked lobster does not store frozen very well. The spiny lobster and crawfish give good results in frozen storage. The North American lobster has serious texture changes when stored frozen. For all crab and lobster species, poor freezing or frozen storage conditions will result in rapid loss of flavor, and the muscle fibers will become dry and tough (texture will be spongy or stringy).

Most shrimp and prawns, both warmwater and coldwater, are frozen to be shipped to market. They can be frozen raw or cooked with the shell on or off. Loss of quality is due mainly to enzymes in the shrimp tissue and to microorganisms from the surface of the shrimp, from the digestive tract, or from the boat deck or the deck hands. Therefore, to obtain the highest quality, it is important to chill as soon after capture as possible and to freeze without a long time in chill storage. Shrimp should be frozen below –20°F (preferably at –25 to –40°F) and stored no higher than 0°F (preferably at –10 to –20°F). Common frozen products are

- *Raw (green) shell on tails.*

- *Raw machine peeled and deveined.*

- *IQF cooked whole.*

- *Cooked, peeled, brine-sprayed with 1-2% salt, IQF, and glazed.*

- *Battered or breaded uncooked.*

- *Hand peeled (tail shell on or totally peeled), IQF.*

Sea Cucumbers

Sea cucumbers are preserved by drying or freezing. Drying is a lengthy process, and artificial drying (oven) requires considerable energy. Therefore freezing and frozen storage are used when available (Slutskaya 1973). They can be frozen raw or after having been boiled.

Prior to boiling, sea cucumbers are dressed by cutting the body wall lengthwise and removing the viscera. If left in, the respiratory tree part of the viscera causes a bitter flavor. The body wall is then boiled in seawater for 30 minutes. It is then frozen and can be stored up to 6 months with no noticeable quality loss. If held at 0°F or lower and temperature fluctuations are kept to a minimum, up to 12 months storage can be achieved. Thaw drip is highest in sea cucumbers boiled before freezing. Protein digestibility is highest in boiled then frozen sea cucumbers. Sensory evaluation of sea cucumbers boiled before freezing and those frozen without boiling showed no difference in quality.

Tanaka and Jiang (1977) looked at three prefreezing treatments of sea cucumbers, as follows:

- *Boiled in freshwater.*

- *Boiled in 3% saltwater.*

- *Soaked in 10% sugar solution for 10 minutes (not boiled).*

The samples were then frozen and stored 1, 2, and 3 months. All had good firmness and texture. It was concluded that prefreezing treatment has no major influence on frozen storage quality and that the essential factors in maintaining good quality of frozen product are storage temperature and post-storage treatment.

Sea Urchin Roe

The only edible part of a sea urchin is the gonad. Both male and female gonads make up what is called sea urchin "roe" or uni. Frozen sea urchin roe is considered to be inferior to the fresh product. Freezing and frozen storage can lead to undesirable flavor and texture change. Many years ago, a literature review was done covering prefreezing treatments and freezing methods (Nordin and Kramer 1979).

The two main sea urchin products are

- *Washed, drained, packed in compartmentalized plastic trays; sold fresh or frozen.*

- *Washed, drained, placed on a clamshell, baked; sold fresh or frozen.*

Processing consists of cracking the test (shell), removing the five gonad segments, and washing with seawater (or a 3% salt solution). Several treatments have been used (mainly to firm the roe) before the roe is graded and packed for sale. If the roe is to be frozen, these are used as prefreezing treatments. The main ones are

- *Soak in 3%, 5%, or 10% sodium chloride for 10 to 15 minutes.*

- *Soak in magnesium chloride in seawater.*

- *Soak in 0.1% citric acid–0.01% propyl gallate*

- *Immersion in 1 molar KC1 or 1 molar calcium chloride.*

- *Dip or soak in alum solution (several concentrations have been used).*

Freezing methods that give the best results are

- *Spraying with liquid nitrogen.*

- *Air-blast freezing.*

- *Plate freezing.*

It has been recommended that brine freezing not be used. There is also a recommendation against use of contact freezing.

There are reports covering the use of vacuum freezing and freeze-drying of sea urchin roe.

Breaded Seafood Portions

Coated fish sticks are usually made from frozen fillet blocks. This allows regular-shaped portions to be cut with lower wastage.

The batter is basically flour in water with or without added breadcrumbs.

There are two types of batters—adhesive, and tempura or puffing. The adhesive batter may have gums or modified starch added to improved adhesion. The tempura batter has sodium bicarbonate and a leavening acid added, which react to produce carbon dioxide and make a puffed batter after frying. Flash-frying has always been a must for tempura-type products, but now there are quick-setting batters and no need to parfry. Frying lasts only 20 to 30 seconds, so frozen products remain frozen.

If cooked, the product should be cooled before going into the freezer. Always freeze immediately after production to keep quality of fish and breading high. Freezing is usually done in a continuous air-blast freezer using a belt conveyor.

Surimi and Surimi Analogs

Surimi is washed refined fish mince. The technology for production was developed in Japan 800 to 900 years ago. The process removes bones, skin, and viscera. About 60 species have been used to make surimi, but roughly 95% of it is made from Pacific pollock. The mechanically deboned, minced fish flesh is washed with water using a multistage process and then pressed dry. Washing removes fat, blood, pigments, and odorous compounds.

Products manufactured from surimi (called surimi analogs) are made possible due to actin and myosin when heated, forming an actomyosin gel that provides the needed texture. With the addition of color and flavors, the following two types of surimi analogs are produced:

- *Japanese kneaded foods called kamaboko type.*

- *Imitation fish and shellfish analogs like shrimp, scallop, and crabmeat.*

Pacific pollock has poor storage characteristics, both in chill storage and in freezing and frozen storage. Consequently, it is best to make the surimi as soon as is practical, because it can be stored frozen. To protect surimi against freeze denaturation, it is important to freeze rapidly to a low temperature, store at a low temperature, and avoid storage temperature fluctuations. With long frozen storage times, textural changes will lead to toughness and dryness, with a loss of water-holding capacity and gel-forming ability (Moosavi-Nasab et al. 2005). To obtain good frozen storage life, cryoprotectants (which act as anti-denaturants) are added. The most commonly used are sucrose and sorbitol.

❄❄❄❄❄❄❄❄❄

Chapter 3   Freezing Systems: Pulling out the Heat

Heat has to flow from warm to cold. Figure 3-1 shows a relatively warm freezing fish giving up its heat to a cold sink, evaporating refrigerant at –30°F in this example. Two categories of machinery are involved in the freezing process. One is refrigeration—the cold sink absorbing the heat at a controlled low temperature. The other is the freezer, a controlled-environment place where heat is removed from the freezing product. Figure 3-1 shows a freezer, in which heat flows from the fish to the air-blast to the heat sink.

Seafood freezing will typically be one of two types, identified by what kind of machinery controls the cold sink. A **mechanical system** uses a combination of compressors, heat exchangers, and controls to create a low-temperature sink of evaporating refrigerant. On evaporation, the refrigerant vapor is then returned to its liquid form in a closed cycle. In a **cryogenic system**, heat flows to very low-temperature fluid supplied from storage tanks. The fluid then evaporates directly into the atmosphere. Both types have value; both have trade-offs. This chapter describes each type of system and the trade-offs involved.

MECHANICAL REFRIGERATION SYSTEMS

The Refrigeration Cycle

Figure 3-2 gives a picture of the basic mechanical refrigeration cycle. The driver of this presumably well-balanced system is the compressor, which takes cool, low-pressure refrigerant gas returning from the evaporator, compresses it, and discharges it at a higher pressure. The compressed gas is also at a higher temperature, as you would expect if you have pumped up a bicycle tire with a hand pump—the discharge hose gets hot.

Figure 3-1. Heat flows from "Hot" to "Cold" while freezing a fish in this typical blast freezer.

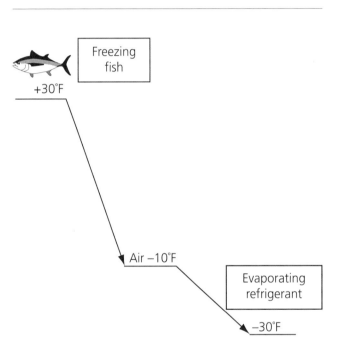

The heat in this high-pressure refrigerant gas is removed when it flows through the coils of the condenser. Onboard a boat, this might be a shell-and-tube or tube-in-tube heat exchanger cooled by seawater. On a shore-based operation, the condenser is more commonly a bank of finned tubing cooled by the outside air and a little water that is blown through by fans.

Removal of heat causes the hot gas to cool, condensing it into a liquid, still at high pressure. The condenser further cools (or **sub cools**) the liquid in order to minimize its chance of suddenly boiling as it flows to the next point in the system, the receiver. The receiver is a storage tank holding

enough refrigerant to ensure that the expansion valve will receive a solid head of liquid.

The expansion valve divides the **high side** (high pressure circuit that includes the condenser) from the **low side** (the circuit that includes the evaporator). It is a valve that is constantly adjusting its opening to let high-pressure liquid squirt through to the downstream low-pressure side. Once the liquid finds itself suddenly under low pressure in the **evaporator**, it begins to boil while absorbing heat.

The downstream coils or piping (inside the evaporator) are kept at a low pressure by the compressor, which continually sucks up and removes

Figure 3-2. Basic refrigeration cycle.

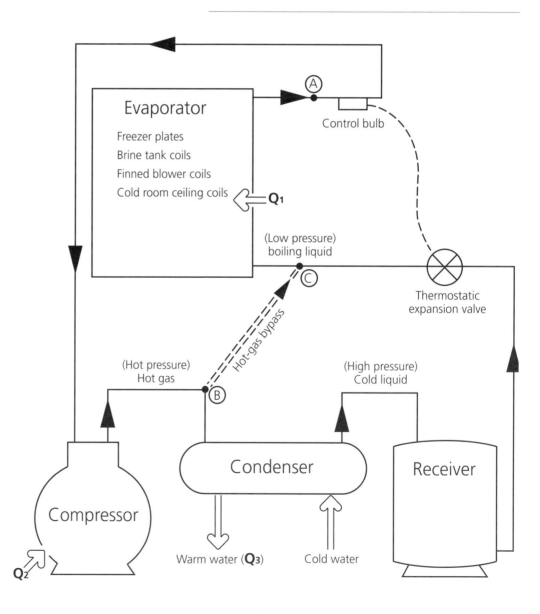

Planning for Seafood Freezing

the refrigerant vapor, sending it back to the high-pressure condenser. Boiling of refrigerant in the evaporator requires heat, the **latent heat of vaporization**, which flows in from the relatively warm fish, air, or brine that surrounds these coils.

If the load (that is, the amount of heat collected per minute) becomes high, perhaps by a few dozen freshly caught salmon laid out on freezer plates, all of the liquid will evaporate before the refrigerant makes it all the way through the coils, and the gas passing point *A* (Figure 3-2) will have warmed up.

This warm temperature is sensed by the **control bulb**, which tells the expansion valve to open up a little more, thus increasing the flow of refrigerant in the system.

As the load decreases (e.g., the fish are beginning to freeze), the evaporator will begin to receive **too much** liquid supplied by the expansion valve. All of it won't boil off in the evaporator and will be mixed with the vapor passing point *A*. The control bulb, sensing a temperature that is too low, will tell the expansion valve to shut down a little. And in this way, the system remains in a dynamic balance.

In any mechanical refrigeration system, there is a basic balance of heat that applies. The major flow rate of heat energy (Q_1 in the figure) comes from the **load**—warm fish or air, for example. This heat is picked up by the boiling refrigerant as it flows through the evaporator.

A second energy flow, much smaller than Q_1, is the work put into the refrigerant by the compressor as it boosts the refrigerant gas to a higher temperature and pressure. The heat equivalent of this work energy flow rate is Q_2 in the figure.

As the refrigerant cools and liquefies in the condenser, the energy that has been added elsewhere in the cycle (Q_1 plus Q_2) must be removed before the cycle can continue. This happens in the condenser; the removed heat is represented by Q_3 in the figure and is sometimes called reject heat. It is always true that Q_3 is effectively equal to the sum of Q_1 and Q_2.

So the expansion valve controls or meters the amount of liquid flowing through the evaporator and responds to the heating load at the time. But it does nothing about the **temperature** in the evaporator. As the load decreases, the expansion valve shuts down, but the compressor keeps going—it keeps sucking vapor out of the coils so that the pressure (called suction pressure) becomes lower and lower.

The refrigerant is in a boiling liquid situation (at saturation), so there is a unique relationship between temperature and pressure—like water boiling right at 212°F when the pressure is right at 1 atmosphere. Therefore, as the suction pressure drops, the suction temperature also gets lower and lower—in the same way that the boiling temperature of water becomes lower than 212°F as you climb in altitude where atmospheric pressure is reduced.

But there are several reasons why you would not want the saturated suction temperature to fall below a certain value. One might be to prevent suction **pressure** from falling below atmospheric pressure; when that happens, any leak that might open up in the piping will allow moisture and contaminants to enter and damage the system. Another reason might be to avoid having a brine or glycol temperature fall below its freezing point. (For properly mixed sodium chloride brine, this would be about –5°F.)

To avoid falling below minimum temperatures, the refrigeration system relies on controls that can act in a variety of ways. One type (a thermostat) senses the temperature of the cooling medium (brine or air). When the minimum temperature is reached, it switches off the power unit driving the compressor.

In some cases, the unit might have some type of pressure regulator that will control evaporator pressure/temperature while the compressor runs continuously. One of these, called a "hot-gas bypass" control, feeds hot gas into the evaporator. This creates a false load and enables the suction pressure/temperature to stabilize at some preset minimum value. In the figure, this controller would act in a flow line installed between points *B* and *C*.

A number of other controls can be built into various systems. Some shut down the compressor in case of abnormal conditions such as too high a head pressure (at the compressor discharge), too low a suction pressure, or abnormally low pressure in the compressor-lubricating oil circuit.

Other controls regulate operating conditions. Besides the expansion valve and hot-gas bypass already discussed, this category includes a water-regulating valve (which regulates cooling water flow to the condenser to control head pressure—the cooler the condenser, the lower the pressure), and a mechanism on some types of compressors that will unload cylinders at decreasing heat loads, effectively decreasing the capacity of the machinery while it is not needed.

Another control might relate to the defrost cycle. In air-blast freezers, moisture carried by the circulating air will condense and solidify on the finned freezer coils. The effect of frost is to act like an insulator, so that the fins can't transfer enough heat. More importantly, it blocks off the air passages, decreasing the flow rate of cold air. Thus it is necessary to periodically defrost, maybe as often as every 2 or 3 hours on some units, depending on loading rate, fin spacing, use of packaging, and other factors.

Defrost systems operate in several ways. One shuts down the compressor and turns on an electric resistance heater mounted in the area of the finned coils in the evaporator. A second closes off the flow of liquid refrigerant while opening a valve in the line between B and C (Figure 3-2), thus introducing hot gas to the evaporator coils. With this technique, the heat melts the frost within 15 or 20 minutes; melt water would run through a hose leading out of the freezer.

You can operate all these defrost cycles manually—that is, you notice a lot of frost on the coils, and you turn the valves or switches yourself. Or the cycles might be activated automatically—by a timer, for example, or by a sensor that detects too high an air pressure drop across the coils.

The simple refrigeration cycle described above is a **dry expansion system** typical of small units. The term means that the liquid phase of the refrigerant is completely gone (evaporated) when it leaves the evaporator (freezer) coils. The leaving refrigerant is dry. An alternate is the **flooded** or **liquid overfeed** system common with larger (particularly ammonia) refrigeration operations. In these, more liquid than needed is pumped through the evaporator coils. As it leaves, the bubbles making up the fraction that did evaporate are separated out, and the liquid is sent around again.

In some cases with both of these systems, the low-side temperatures are very low, requiring a very high lift from the evaporator temperature to the condensing temperature. To maintain efficiency and performance, this usually calls for a two-stage system—one warmer refrigeration system taking heat from the colder low-stage cycle.

About Refrigerants

Until recent years, the smaller low-temperature refrigeration systems, as well as most systems onboard vessels, commonly used either R-502 or R-22. These are among a series of compounds called **halocarbons**, which users still commonly refer to as **freons**. This term, which took on a generic meaning for halocarbons, was actually a trade name of the Dupont Chemical Co, one of the big manufacturers. Ammonia (also labeled R-717) is a naturally occurring substance and has long been used in larger, usually shore-based systems.

The two halocarbons (R-502 and R-22) in common use for low-temperature refrigeration both contain chlorine in their molecular structure. And both break down in the atmosphere, releasing chlorine into the upper atmosphere. R-502 is one of a class called chlorofluorocarbons (CFCs) that releases the most chlorine. With the recognition that these released chlorine atoms create serious environmental damage, most countries (including the United States in 1995) banned the manufacture of R-502 along with all refrigerants in the CFC category.

One group of replacements for R-502 will come from a category of compounds called **hydrofluorocarbons** (HFCs) whose molecules contain no atoms of chlorine. They are thus environmentally safer in that regard. Two HFCs that appear to be common industrial substitutes are R-507 and R-404a. These both have different pressure-temperature characteristics. R-404a is actually a blend of compounds, such that evaporation and condensation do not occur at the same exact temperature.

The other common halocarbon, R-22, is in yet a third class of compounds called **hydrochlorofluorocarbons** (HCFCs). They will also break down and release some chlorine in the atmosphere, but to a lesser extent than R-502. The use of R-22 is being diminished worldwide, with estimates that U.S. production will cease within the next 20 years or so. However, at the present time, R-22 remains a major contender as a low-temperature refrigerant.

With all of the environmental and economic costs of these manufactured synthetic refrigerants, many question the use of them at all. They advocate use of a "natural" refrigerant, the most common of which is ammonia. It is relatively cheap, has superior levels of energy efficiency, is environmentally harmless, and is easily detected when leaking. But with high operating pressures, it requires heavier (and so, more expensive) machinery—steel (vs. copper) piping is required. Operation and servicing require special training. It can be toxic above a certain concentration and explosive above another, so safety measures are important. Thus, ammonia systems are currently cost-effective only above a certain refrigeration capacity. Work continues, particularly in Europe, on the use of natural refrigerants: ammonia, carbon dioxide, natural gas, and others. Ammonia is beginning to find applications in smaller systems. And two-stage systems (carbon dioxide in the low stage, ammonia in the high stage) are being used to support very low-temperature mechanical refrigeration.

Refrigeration Terms

Some of the terms that you may find in refrigeration literature deserve a bit more explanation (see also the definitions in the Appendix).

Btu means British thermal unit. It is the amount of heat involved in changing the temperature of a pound of water by 1°F. This unit of heat is used in the English system but not in the metric system, where the terms **Calories** and **Joules** are more often seen. **Btuh** means Btus per hour.

Ton (or **refrigeration ton**) is a rate of heat flow. It commonly refers to the amount of heat that can be removed by a freezer or refrigeration system within a certain period of time. **One ton** equals 12,000 Btuh, and is the rate of heat absorbed by a ton of ice that melts in 24 hours.

Capacity is the amount of heat (per unit of time) the freezer or refrigeration system will remove. It varies greatly with suction temperature, which in turn varies with the air, brine, or plate temperature you wish to maintain. Figure 3-3 shows a representative capacity curve. If someone says a compressor has a 12-ton capacity, you should ask, "At what suction temperature?" Capacity of a compressor or system is frequently quoted for a suction temperature of +40°F. But a capacity of 12 tons at +40°F suction temperature translates to something less than 2 tons when temperatures fall into the blast freezer range (Figure 3-3). The reason is that gas boiling off at +40°F is much more dense than

gas boiling off at lower temperatures, say –25°F. At either level, the piston compressor continues to pump at a constant rate of displacement (in **CFM**, or cubic feet per minute). So in a given period of time, much more of the higher-density refrigerant gas, i.e., that boiling at 40°F, will be pumped around the system. A greater flow rate of refrigerant means a higher capacity. (A confusion that might arise while troubleshooting involves the terms **capacity** and **temperature level**. If you observe that a system can pull down to a very low temperature while under a low load, you can't then assume that all is in order. The important question is "How **fast** does it pull down with a known load?")

Horsepower often refers to the power needed to drive the compressor. In this case, it would be the required output of a hydraulic motor, electric motor, or diesel engine. At warm temperatures (say, +40°F), the rated **tons** and **horsepower** numbers for a given system will be fairly close.

Secondary refrigerant refers to a liquid such as a brine or glycol that carries heat from the fish to the primary refrigerant flowing through the compressor. Freezing contact plates might have a secondary refrigerant flowing through their internal channels. In some systems, the brine immersion tank might be chilled by a glycol secondary refrigerant flowing from the refrigeration unit to coils lining the tank walls.

Heating load results from warm fish and from other sources. Some of these are

- *Heat leaking through the walls of the freezer.*

- *Heat given off by people working in the room of a large freezer.*

- *Defrost cycle in the case of blast freezers. Don't make it any longer than necessary.*

- *Inflow of warm air from an open or poorly fitting door.*

- *Fan energy. All the power that goes into moving air ends up as heat in the air. That's why manufacturers don't use fans that are larger than necessary.*

- *Pump energy as in a glycol or brine system.*

For some blast freezers, 50% of the heating load might be caused by factors other than warm fish.

The freezers that follow typically operate with a heat sink provided by mechanical refrigeration.

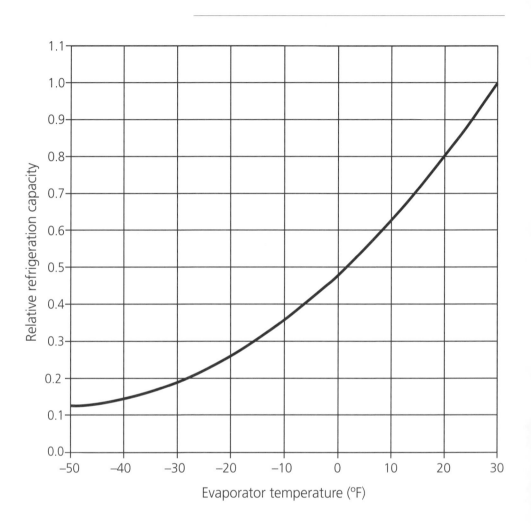

Absorption Refrigeration

Although not yet common in seafood freezing, absorption refrigeration represents a variation on the *mechanical refrigeration* theme. Both circulate refrigerant in a closed loop, and both have evaporators in which the liquid refrigerant absorbs heat as it turns into a gas. The difference lies in the compressor, which pumps low-pressure vapor to the high pressure side. Mechanical systems use a motor-driven mechanical compressor, as in the process just described. Absorption systems use *molecular forces* and *thermal energy* to do the same thing—essentially a *thermal compressor* (Dorgan et al. 1995, Kallenberg 2003). The low-temperature gas mixes with an absorbent liquid, creating a solution. This liquid solution is pumped to a high pressure with much less mechanical energy than is required for a mechanical compression of the refrigerant gas. The solution is then heated, breaking the molecular bonds between the refrigerant and absorbent liquid. This releases the refrigerant gas to the high-pressure side, after which it is condensed to a liquid, then expanded, the same as in the mechanical refrigeration system. For low-temperature (freezing) processes, the refrigerant is commonly ammonia and the liquid is water.

One advantage of absorption refrigeration emerges when the required heat energy is "free." This might be waste heat from an industrial process, geothermal or solar energy, or engine heat given off in diesel-driven generator plants common in western Alaska communities.

Feasibility of these systems will depend on the temperature and capacity of the heat source, and on the required evaporator temperature. Using engine jacket water at around 165°F, Energy Concepts Co. (1994) designed and installed an ice-making plant in Kotzebue, Alaska. This demonstration plant reliably produced 10 tons of ice per day using a patented multistage absorption cycle (Erickson 1996).

There is continuing interest among several Alaska communities in absorption technology to refrigerate community cold-storage warehouses. With ongoing developments, these could make use of both engine jacket water heat and the higher-temperature exhaust heat (~650-850°F) as the major sources of driving energy. With quiet operation, and expected low-maintenance, such cheap-energy systems remain a potential that should be continually evaluated.

Airflow Freezers

Three different air-blast systems might be typically used for freezing seafoods: spiral, tunnel, and blast freezers. The type of freezer is defined by the nature of the cabinet where air and product come in contact. In each system, the refrigeration machinery pumps liquid—whether ammonia or halocarbons—to an evaporator likely made up of banks of finned tubing. There, powerful fans blow air through the fins; inside the tubing, refrigerant boils off, absorbing heat from the recirculating air. The cold air then heads off to collect more heat from the freezing product. Air temperatures would ideally approach –40°F; however, there are some recent onboard systems supporting much colder air temperature freezing albacore tuna for high-value markets.

Unless the products are packaged, all of these systems must be frequently defrosted, because any moisture lost from freezing products will deposit as frost on the colder coil surfaces. Frost will quickly restrict the amount of airflow. Defrost cycles will depend on the product (wet or packaged) and freezer design, and may be anywhere from once a week to once every freeze cycle. In some blast freezers with wet fish, the operators have scheduled a defrost about an hour after freezing begins. At this point, the evaporation of water from the crust-frozen surface of the fish is greatly reduced. When free of frost, the defrosted fins allow the evaporator to operate at a high efficiency for the remainder of the freezing period.

Spirals

As the name implies, product enters this freezer on a belt, which travels in a spiral motion through a near-cube-shaped room (Figure 3-4). The airflow direction, depending on the design, is horizontal, vertical, or some combination of those, as it flows over the product riding along on the belt. Both the belt speed and loading density can be controlled to ensure complete freezing. Most manufacturers also include an option for some kind of continuous belt-cleaner outside the freezer.

Spirals enable a continuous-feed, well-controlled freezing process at very low air temperatures (typically approaching –40°F), supporting a variety of product shapes and sizes. They also occupy a relatively small amount of floor space in the plant.

Tunnels

The tunnel freezer, in its simplest form, is a straight, continuous link-belt carrying product through a tunnel. There it is pummeled with high-velocity cold air supplied by evaporator/fan units housed alongside (Figure 3-5). Different manufacturers present a number of variations to this theme. For example, airflow can be horizontal or vertical. The product's route through the tunnel can also be made up of several passes to prolong the time in the freezer. Different belts can even operate at different speeds; for example, wet product is loaded in a single low-density layer on the first belt to create a quickly frozen outer crust. This forms a good moisture barrier. It then dumps the product onto a second, slower belt where it piles up into multiple layers while completing its required freeze time.

In a **fluidized bed** variation, small, individually quick-frozen (IQF) products like shrimp, surimi seafood chunks, peas, and berries riding on this second belt are literally suspended in a blast of cold air directed upward through belt and product mass. This turbulent suspension of particles knocking against each other makes for a very high rate of heat transfer. And in another variation, some tunnels accommodate wet or runny products that are first crust-frozen on a solid belt having cold contact from below; it then completes the freezing on a second pass through the tunnel.

Tunnels perform the same function as spirals but take up more floor space. They also tend to be mechanically simpler and require perhaps less attention to adjustment and maintenance. Both tunnels and spirals are well-suited to small items having short freezing times of less than an hour or so. However, large systems can be designed for high flow rates and freezing times of several hours.

Blast Freezers

The term **blast freezer** commonly refers to a batch freezing operation in which the product, loaded on trolleys, is wheeled into a room or large cabinet where it remains until frozen. Blast freezers are sometimes called **batch-continuous** systems if the trolleys are periodically removed row-by-row on a "first-in-first-out" basis. In the room, powerful fans blow circulating air through evaporators located above a false ceiling (Figure 3-6). This cold air then flows uniformly over the product loaded on the

Figure 3-4. A spiral blast freezer with air flow vertically downward through the belt.

(courtesy Northfield Freezing Systems)

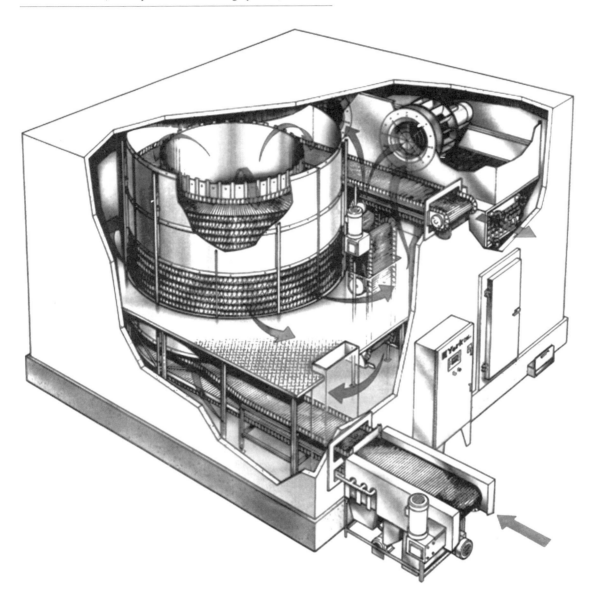

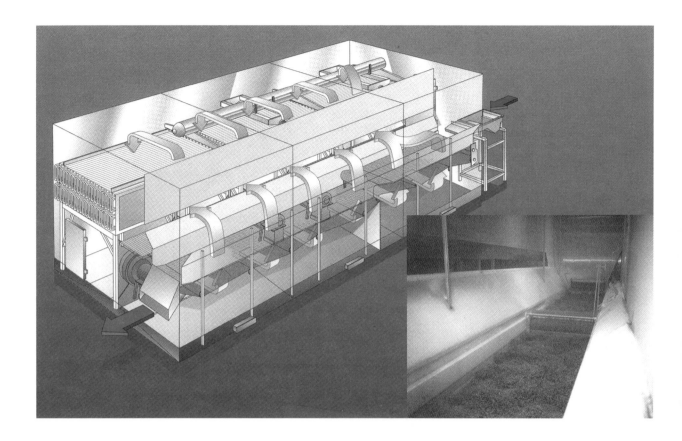

Figure 3-6. A batch blast freezer requires uniform air flow through the trolleys.

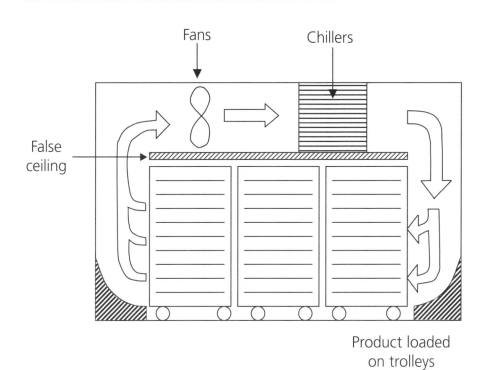

Fans

Chillers

False ceiling

Product loaded on trolleys

racks or trolleys below. Large products that take a long time to freeze are more suited to a batch process than to the continuous process of a spiral or tunnel.

Assuming proper design, a major impediment to performance has to do with product loading and uniform airflow. Without a balanced resistance and/or proper vanes to direct air as it first hits the trolleys, flow velocities over products in different locations will vary significantly. As shown in Chapter 1 (Figure 1-8), such a variation can seriously affect freezing time. Figure 3-7 depicts one of these imbalance problems: without a uniform resistance to the air over the cross-section of the room, the air will seek the shortcut, or path-of-least-resistance, resulting in imbalanced velocities. A second flow problem can result when the air turning the first corner and hitting the trolley is not properly directed. Figure 3-8 shows how airflow in one blast freezer is distributed over a load represented by just one empty trolley.

Brine Freezers

Brines typically refer to salt solutions, but glycols and other fluids fall into this category as well. These liquids, after flowing over the freezing product, circulate through a heat exchanger where they give up heat to the vaporizing refrigerant. The cold brine (sometimes referred to as a **secondary refrigerant** or **secondary coolant**) is then pumped back to sprayers or immersion tanks where it removes heat from the freezing product. The product might be wrapped in a sealed package to prevent direct contact of the brine with the food. When brine directly contacts packages, some post-freeze cleaning would presumably be needed. One processor rinsed brine-frozen packages in freshwater, then whacked them to knock off the resulting ice, before packing into cartons. A common exception to the use of packaging is onboard tuna boats, and on some shrimp boats in southern waters, where whole fish or shrimp freeze in direct contact with refrigerated sodium chloride brine.

Figure 3-7. Eliminating the chance for air-bypass is important for good blast freezer performance.

(adapted from Graham 1974)

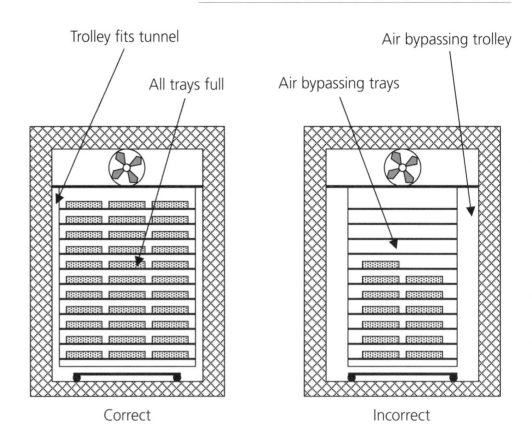

Trolley fits tunnel

All trays full

Air bypassing trolley

Air bypassing trays

Correct

Incorrect

The freezer can consist of a box or fish-hold in which a series of spray heads distribute brine over warm product lying below. In some older tunnel freezers, sprayers direct brine onto the bottom of a sheet-metal conveyor carrying unpackaged product. Many spray-brine systems require anti-foaming agents.

The freezer can also consist of a tank in which circulated cold brine envelops the fish or packaged product. Agitation will greatly increase the rate of heat transfer. Fikiin (2003) reports recent European progress with a method labeled HydroFluidization. Similar to the airflow tunnel described above, jets of brine shoot upward through the product, suspending it in the turbulent bath. Using sodium chloride brine with typical temperatures of 3°F, small fish were frozen exceptionally fast. Experiments showed even greater rates of freezing with

pumped ice slurries of the type supplied by a number of commercial suppliers. Made from brines using a variety of solutes, such slurries can carry other beneficial components such as antioxidants or cryoprotectants.

A major advantage of brine freezers over air-blast freezers is that they can operate with higher energy efficiency. This relates to good heat transfer at a much higher refrigeration operating temperature. **Heat transfer coefficient** is a term that describes the relative ease with which heat flows from the freezing product surface to the surrounding cold medium. In brines and other liquids, this coefficient is higher than in most air-blast systems. As a result, it can be shown that for some product sizes, freezing can be about as rapid in typical brines (e.g., 0°F), as in the much colder (e.g., –40°F) air-blast systems (Kolbe et al. 2004a). And this

Figure 3-8. Unbalanced air flow pattern in a lightly loaded blast freezer. The profile shows air velocity, in feet-per-minute.

(adapted from Kolbe and Cooper 1989)

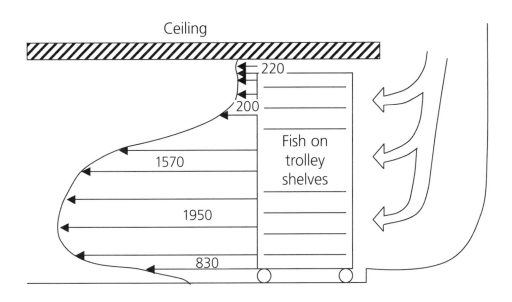

means that the refrigerant evaporating temperatures, and thus the energy efficiency, will be higher (Figure 3-3).

Common brines and fluids are described here.

Sodium Chloride (NaCl)

Mixing sodium chloride salt with water at a weight ratio of 23.3% results in the "eutectic point" of –6°F, the lowest freezing temperature you can achieve with this brine (Hilderbrand 1979). It is not practical, however, to operate the freezer at a temperature this low, due in part to the temperature drop that must exist between brine and the boiling refrigerant in the heat exchanger. Attempts to do so would cause brine to begin freezing on the relatively lower-temperature tube walls. So a practical minimum operating temperature for sodium chloride brine might be around 0°F.

Calcium Chloride (CaCl₂)

The theoretical eutectic point of this salt brine is –67°F at a solution of 29.9% by weight (ASHRAE 1997). Thus calcium chloride brine can operate at a lower temperature than sodium chloride without

the same risk of freeze-up. However, the brine becomes quite viscous at such low temperatures, making it hard to pump; handbooks indicate minimum practical application temperature closer to –4°F. In addition, unlike sodium chloride, calcium chloride in contact with food is considered problematic, if not toxic.

Glycols

Both ethylene glycol and propylene glycol are used as secondary refrigerants, although the former is shunned for food applications because of its greater toxicity. Its advantage is that it is less viscous than propylene glycol at low temperatures. Both may have added **corrosion inhibitors**, as do some of the salt brines, and this creates some food quality problems as well as environmental or treatment plant problems regarding disposal. When mixed with water, the freezing point of propylene glycol can be as low as –58°F. However, as with calcium chloride brine, the fluid is extremely viscous at such low temperatures, demanding high pumping energy and displaying diminished heat transfer. A 0°F operating temperature at a mixture of roughly

50:50 is a minimum temperature target (ASHRAE 1994, 1997).

Other Brines and Fluids

A few manufacturers advertise proprietary formulations of heat transfer fluids and organic solvents that can be used in brine freezers. Based on petroleum, silicone oil, citrus oil, or other compounds, these fluids can operate easily at –40°F. One example is "Isopar H," a product of Exxon Mobil Chemical. Another is Paratherm LR, with a usable temperature range of –40°F to +400°F (Paratherm Corp. West Conshohocken, Pennsylvania). Although Paratherm LR is expensive ($1,100 for a 55-gallon drum), such cold temperatures, combined with any agitation at all, would ensure a very rapid freezing rate of small, individual packages.

Contact Freezers

Horizontal Plate Freezers

Horizontal plate freezers are well suited for blocks of fillets, mince, or surimi, and for packages having predictable dimensions, rectangular shapes, and large, flat surfaces. Using surimi as an example, a typical 10 kg (22 pound) batch is metered into 2 or 3 mil polyethylene bags, with final block dimensions on the order of 2.25 × 12 × 23 inches. The end of the bag is then folded under the block, which is placed within an aluminum or stainless steel tray. After loading onto the plate freezer shelves, hydraulic rams press the upper shelf hard onto the package, with appropriate spacers to prevent crushing and to ensure uniform thickness of the frozen package (Figure 3-9). Some products, such as prepackaged cartons, are loaded typically with a single aluminum surrounding frame. This allows good contact and heat transfer while controlling the frozen package thickness.

Note that both under-filling and overfilling can create difficulties in the process. The former allows an air gap and poor heat transfer at the upper surface of an individual pan; the latter may lead to package distortion or tearing and possible air gaps at the top of adjacent packages on the plate. A buildup of ice on the plates can lead to poor performance as well (Figure 3-10). The consequence of these uneven contacts is described in Chapter 1 (Figures 1-16 and 1-17).

The moveable shelves in a typical plate freezer are most commonly made of extruded aluminum, with internal channels to pass refrigerant from one side of the plate to the other (Figure 3-9). Because plates move down and up as the operator loads and then unloads blocks, refrigerant must be supplied to the plates through flexible hoses. An older plate design consists of a serpentine grid of square steel tubing, housed within two parallel steel sheets making up the surfaces of the plate. An outer border seals this thin box; a vacuum and filler material ensures good heat transfer between tubing and the outer plate. The steel plates tend to be on the order of 15% less expensive than those made of extruded aluminum. However, the one-piece aluminum structure ensures a better transfer of heat from the outer plate surface to the refrigerant flowing within the passages. So freezing with the aluminum plates is faster.

The refrigerant used in most shore-based plate freezers is ammonia, which floods the plates from reservoirs installed outside of the freezer enclosure. Heat flows from the freezing product to the relatively colder refrigerant, which continually vaporizes at a low pressure. The liquid/vapor mixture in this flooded system then flows either by gravity (called a **flooded system**) or by means of a pump (in a **liquid overfeed system**) to one of the external reservoirs where it separates. Liquid goes back through the freezer plates; vapor is pumped to the compressor where it is pressurized (to the high side), liquefied in the condenser, and returned to the low-pressure reservoir.

The halocarbon refrigerant R-22 is often used in liquid overfeed plate freezers on factory ships. This is to avoid some safety and regulatory complexities that exist with ammonia. A downside is that the per-pound cost of R-22 is greater than that of ammonia by a factor of about seven. This is significant for systems holding large volumes of refrigerant.

Halocarbons are common in small systems that use these refrigerants in a self-contained compressor/condenser unit requiring a minimum volume of refrigerant to operate. The refrigerants are injected into the plates through a restriction valve, called a dry expansion (vs. flooded, or liquid overfeed) supply system. This means that high-pressure liquid is metered into the low-pressure plate passages by a thermostatically controlled expansion valve. The expansion valve enables only

Figure 3-9. A horizontal plate freezer shown without a cabinet enclosure. The inset shows a cutaway of an extruded aluminum plate.

(courtesy Dole Refrigerating Co.)

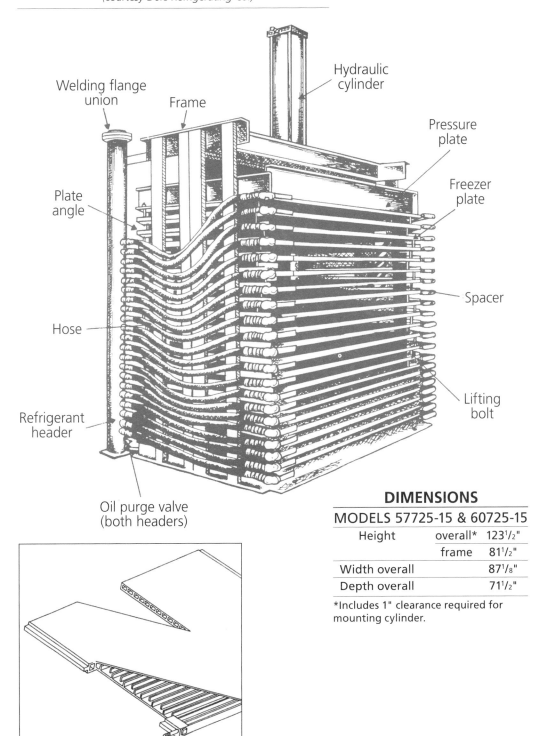

Welding flange union

Frame

Hydraulic cylinder

Pressure plate

Plate angle

Freezer plate

Hose

Spacer

Refrigerant header

Lifting bolt

Oil purge valve (both headers)

Aluminum plate cutaway view

DIMENSIONS

MODELS 57725-15 & 60725-15

Height	overall*	123½"
	frame	81½"
Width overall		87⅛"
Depth overall		71½"

*Includes 1" clearance required for mounting cylinder.

Figure 3-10. Some sources of poor performance in a horizontal plate freezer.

(from Johnston et al. 1994)

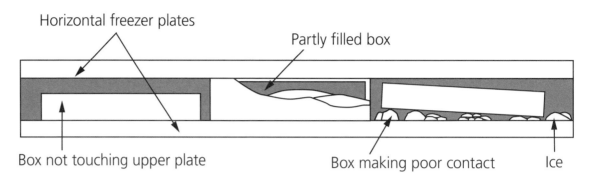

Horizontal freezer plates

Partly filled box

Box not touching upper plate

Box making poor contact

Ice

enough liquid needed to absorb the freezing load at that moment. Thus the liquid becomes completely vaporized shortly before reaching the end of the passage, where it is then sucked up by the compressor. A major disadvantage of a dry expansion system is that temperature and heat transfer conditions inside the plates tend to vary throughout the freezing cycle. Initially the thermostatic expansion valve cannot supply an adequate flow rate, and the temperature tends to increase. But in a liquid over-feed system, pumps continually supply three to four times the refrigerant needed to absorb the heat load. The temperatures and conditions therefore remain relatively uniform (ASHRAE 1994).

The cold refrigerant temperature inside the plates is uniquely related to the vapor pressure, because the mixture is **saturated** (i.e., liquid and vapor exist in equilibrium). Thus, one can indirectly measure the boiling refrigerant **temperature** (called the **saturated suction temperature**) by reading the suction **pressure** on a gage, and then finding the unique saturation conditions for that particular refrigerant. In virtually all cases, however, the suction-pressure gage will have the corresponding saturation temperatures printed on its face for the particular refrigerant used in that freezer. During product freezing, there will be a few degrees of temperature difference between the plate surface and the flowing refrigerant. But by reading the suction-pressure gage, you know the approximate plate temperature.

In addition to ammonia and halocarbon refrigerants, a third type, less common for these plate freezers, is a secondary refrigerant such as propylene glycol. It is refrigerated in a heat exchanger somewhere outside the freezer, then pumped through the plates. A somewhat uniform passage temperature can be maintained if the flow rate is sufficiently high. The advantage of this system is that potential leaks (e.g., at the flexible supply hoses) present less risk. The disadvantage is the increased energy consumption caused by additional pumping and by the lower (and thus, less efficient) evaporating refrigerant temperature in the external heat exchanger, which cools the secondary refrigerant.

And finally, carbon dioxide is beginning to find use as a refrigerant applied to plate freezers by at least one European manufacturer. Besides being a "natural" and environmentally safe refrigerant, it enables exceptionally low plate temperatures and fast freezing rates.

Other Types of Contact Freezers

While horizontal plate freezers are the most common of the contact category, there are others that may be perfect for the job.

Vertical Plate Freezers

The same extruded aluminum freezer plates, installed vertically, make up this freezer. Developed about 50 years ago, it enabled the British distant water fleet to quickly freeze cod for storage and

Figure 3-11. A vertical plate freezer. Frozen blocks are commonly ejected by rams pushing up from below. Filling slots with meat offal in this example is by hose at the left.

(courtesy DSI-Samifi)

later processing ashore. Besides dressed fish, other options such as mince or semi-liquid products can be pumped into the vertical space between plates. Once freezing is accomplished, hot gas is pumped into the plates, allowing them to be pulled apart. Commonly, rams eject the frozen blocks out the top (Figure 3-11).

Shelf Freezers

If freezer plates are fixed as stationary shelves, the system becomes far less complex (and so, less expensive) than the horizontal plate freezer. Freezing times are likely longer, because the only contact on the top of the product is with cold air. At least one manufacturer produces an extruded freezer plate that has a finned lower surface (Figure 3-12). When housed within a cabinet, these fins produce good heat transfer between air and plate, rapid freezing rates, and the option of freezing different-shaped/sized products.

One further configuration was developed by the Vancouver Technology Lab in the 1960s. Called a **combination freezer**, it had a blast freezer unit blowing cold air over a set of freezer plate shelving. Performance was quite good (Gibbard 1978).

Contact Belt Freezers

In these tunnel freezers, fluid or wet products are placed on a moving stainless steel belt. As the belt leaves the tunnel and curves over the exit roller, the frozen product, e.g., fish fillets, pop off and fall into a collector or conveyor. The equipment may include a washer/cleaner to prepare a hygienic belt before reloading with wet product.

Refrigeration comes in several forms. An air-blast may be most common, cryogenics such as liquid nitrogen or carbon dioxide can be used, and some older freezers operated with cold brine sprayed onto the bottom surface.

Figure 3-12. **A shelf freezer having the refrigeration unit contained in the cabinet below. The extruded plate cross-section shows fins lining the lower surface, cooling air, which then flows over the product below.**

(courtesy Gunthela Enterprise, Ltd.)

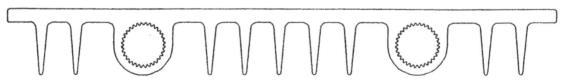

Extruded plate

CRYOGENIC REFRIGERATION SYSTEMS

The term **cryogenic** simply means very low temperatures. In food freezing, it refers almost exclusively to the use of liquid nitrogen and liquid carbon dioxide, fluids that are delivered to the plant by tanker truck. Cryogenic refrigeration differs dramatically from mechanical refrigeration. The cold sink is the cryogenic refrigerant, sprayed or pumped directly into the freezing cabinet or tank. As it collects heat, it vaporizes and eventually discharges to the atmosphere. Temperatures can be very low. Liquid nitrogen at atmospheric pressure vaporizes at –320°F; liquid carbon dioxide first turns to "snow" when vented to atmosphere, and then it vaporizes at –108°F. As a result of these low temperatures, freezing is very rapid and under some conditions can be too fast. For example, a whole fish fillet dropped into a liquid nitrogen bath would likely crack into several pieces.

Liquid Nitrogen (N₂)

A plant using a liquid nitrogen (LN) freezing system would need to erect a holding tank while setting up a contract with an LN supplier. The supply must be reliable and close enough so that delivery costs are feasible. One supplier noted that about 150-200 miles by road might be considered a maximum operating distance. Most LN companies can set up an automatic monitoring-and-refill option. Storage tank size will depend on the plant's processing rate and proximity to the supplier. Common capacities range from 900 to 12,000 gallons. A rough rule-of-thumb: each pound of fish frozen will use around 1.2 pounds of LN (Persson and Löndahl 1993); this will increase to 1.5 pounds (or 0.22 gallons) with equipment cool-down, leakage, and poor insulation on storage tanks and piping. It can be as high as 2.0 (Löndahl 1992, ASHRAE 1994).

The LN tank is designed for 20-250 psi of pressure; it will be well-insulated to minimize heat leakage from the outside. But the temperature difference between LN and outside air is extremely large, and a perfect insulator is not available. So some small but continuous rate of evaporation and loss will occur. Users report losses that can exceed 1% per day—a "use it or lose it" scenario. And for small tanks, this would be higher.

Additional losses will occur by heat leakage into the pipes carrying LN to the freezer, so minimizing that distance and using good insulation are significant. Commonly, LN piping will be surrounded by a very thick layer of foam insulation. But even this will allow heat to leak in and vaporize the –320°F fluid. Contractors have suggested that for runs longer than 100 feet or so, vacuum-jacketing may be cost effective (Weiner 2003). In one contractor's example, heat penetration with normal (urethane) insulation was on the order of 20 Btu per hour per foot (Dewey 2001). This would mean that over a 200-foot distance between tank and freezer, 7 gallons would vaporize and be lost during each hour of operation. Static vacuum piping is said to reduce heat leakage by a factor of 40, but it costs almost twice as much to install—another planning trade-off (Weiner 2003, Whited 2005).

Liquid Carbon Dioxide (CO₂)

As liquid CO_2 vents into the freezer cabinet, it converts first to a solid/vapor mixture that appears as "snow." (If compressed into blocks, this snow would become the solid dry ice we're all familiar with.) The solid snow then converts to a vapor while absorbing its heat of sublimation—the heat pulled from the freezing product. The cold vapor may then collect additional heat before it exhausts to the atmosphere. The rate of use for liquid CO_2 is roughly the same as that for LN—somewhere in the range of 1.2-2.0 pounds per pound of frozen product. As with liquid nitrogen, its efficiency of use will depend on the "overhead" losses. For example, in a small cabinet batch freezer, these losses will be much higher than in a continuous-flowing tunnel freezer (Lindsey 2005).

Typical liquid CO_2-insulated storage tanks range in capacity from 3,500 to 12,000 gallons and store 0°F liquid at 300 psi. (Liquid CO_2 density is 63.4 pounds per cubic feet; 1 gallon has a mass of 8.5 pounds.) Because the storage temperature is so much higher than that of LN, losses caused by heat leakage can be controlled to zero. This is aided in hot-weather/low-use periods by a small mechanical refrigeration unit that will switch on to refrigerate coils lining the tank walls. As with LN, the supply company can provide full service, remote monitoring and troubleshooting, and timely refill as needed.

Freezers

Many of the same blast freezers described in the previous section can be adapted for cryogenic refrigerants. Vaporizing cryogenic fluids swirling around products replace the swirling air that was cooled in separate finned-coil heat exchangers. Cryogenic systems are particularly good at freezing smaller products in which heat can rapidly travel from the core to the surface.

One typical LN freezer is designed as a tunnel, with LN spray heads located near the exit end (Figure 3-13). As the evaporating liquid spray absorbs heat from the freezing product, fans blow the cold vapor in a direction counter to that of the conveyed product. This enables the vapors to rapidly pre-cool and crust-freeze the entering product. Consequently, the tunnel design allows for efficient use of the refrigerant and avoids radical temperature changes and stresses that might cause the product to crack.

Both LN and liquid CO_2 are used in spiral freezers (Figure 3-4) with large fans but without the air-to-refrigerant heat exchangers. For small lots, cabinet freezers like that shown in Figure 3-14 employ both LN and CO_2; fans are used to maintain high gas velocity over the product surface. Another variation common with CO_2 freezing small food items uses a rotating drum that directly mixes the product and cold gas.

As stated, large products will crack if immersed directly into a pool of LN. However, at least one West Coast processor has found a small-bath system effective in IQF freezing of cooked/peeled pink shrimp. A wheel flings shrimp into the LN bath; the individual shrimp immediately crust-freeze. The shrimp are removed so rapidly from the bath that the core is still unfrozen, left to solidify, stress-free as the shrimp move on to the glaze/pack-up station.

Although not common, one processor has used LN to cool brine to freeze cooked crab in a large sodium chloride brine tank. A metered stream of LN fed directly into a stream of circulated brine maintained the required low operating temperature in the tank.

Figure 3-13. Liquid nitrogen tunnel freezer. Spray injected at the product exit end vaporizes and flows toward the product inlet end

(Courtesy Air Products and Chemicals, Inc.)

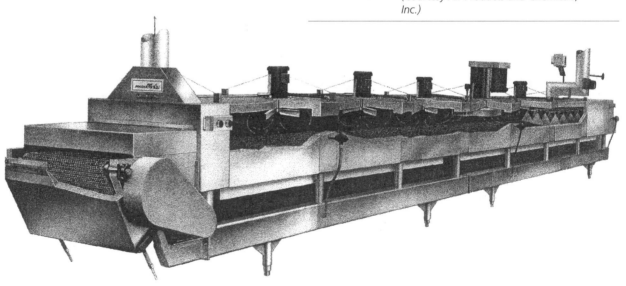

Figure 3-14. Liquid nitrogen cabinet freezer

(courtesy Martin/Baron, Inc.)

TRADE-OFFS: MECHANICAL VS. CRYOGENIC REFRIGERATION

The choice between mechanical and cryogenic refrigeration could depend on what is already in the plant. Is a system expansion easier/cheaper than installing a new refrigeration system from scratch? If an existing system is not an influence and you do start from scratch, which is better?

The introduction to this manual lists some critical information needed to formulate a plan. When making the most common comparison—cryogenics vs. air-blast freezing—several issues are significant.

Required Freezing Rate

Chapter 2 describes the influence of freezing rates on food quality. With the very low temperatures of LN and CO_2, cryogenics can freeze your product very fast. Well-designed and operated air-blast freezers, too, can produce fast freezing rates. Take care when using the "faster-is-better" criterion. Measure the frozen food quality after a representative period of cold storage at anticipated storage conditions. Food technologists find that seafood stored at typical commercial conditions can quickly lose any benefits of exceptionally fast freezing.

Dehydration Loss

With the exceptionally high initial freezing rate of cryogenics, the early formation of a frozen crust can limit moisture loss during freezing to a fraction of a percent of weight. By contrast, moisture loss in an air-blast freezer might be on the order of 1-2%, and in one that is poorly controlled, up to 4-5%.

To take advantage of the very rapid crust freezing of cryogenics, coupled with a relatively lower operating cost of mechanical refrigeration, systems have been created that combine the two. Such a cryomechanical freezer might combine LN followed by air-blast in a tunnel configuration for unpackaged products.

Costs

The conventional wisdom is that when cryogenics are compared to mechanical refrigeration, the relative capital costs of cryogenic systems are lower and the operating costs are higher.

Mechanical systems, particularly those using ammonia, require an engine room full of expensive machinery, with trained operators and regular maintenance. But this machinery is powered by electricity, which, at least in the Pacific Northwest, is fairly cheap.

Cryogenic refrigeration equipment includes a tank, insulated piping, and some controls. The tank is often leased from the gas supplier. Therefore, a major operating expense becomes the cost of the delivered LN or CO_2. Trucking costs will vary with distance from the supplier. In the case of LN, more than 1% is lost each day due to heat leakage into the tank. The shut-down of a line, or interruption of fish delivered, will use up some liquid nitrogen whether it is freezing or not.

Cost analyses that compare these two systems have to be based on each specific process. And one major assumption is the product weight-loss expected for each. One such analysis by a cryogenics vendor considered freezing hamburger patties and assumed losses of 1% into the blast freezer; 0.3% in the LN freezer (Elenbaas 1989). With other assumptions thrown in, and weight losses equated to lost product revenue, the results showed freezing cost (in dollars per pound) were similar. Would a lower rate of dehydration also produce a higher-valued product? It is impossible to make such claims without measuring both quality effects and dehydration weight losses on your own product. Vendors of both systems commonly invite potential customers to bring some product to a demonstration plant to observe effects.

Air-Impingement Freezers

A new development in air-blast freezing, particularly in a few commercial tunnel freezers, uses air impingement jets to shoot high-velocity air directly onto the product surface. With velocities on the order of seven times those in well-designed conventional blast freezers, freezing times for thin products are said to approach those in cryogenic freezing. Salvadori and Mascheroni (2002) report experiments for hamburger patties that showed weight loss was also significantly lower than in conventional blast freezers; much of this was a result of the shorter freezing time.

Flexibility

Short-term surges in production rates might be handled more easily with cryogenics. An added extension to a freezer tunnel, or an added cabinet, could be refrigerated essentially by cranking up the rate of gas flow.

Gas Supply

The entire freezing process depends on gas supply. How far must the supply of LN or CO_2 be trucked? How reliable is the supply?

Chapter 4   System Selection and Layout

Previous chapters have begun to outline the planning steps leading to a new freezing line. The design engineer, refrigeration contractor, or freezer supplier would fill out these and many other steps with the necessary details, based on analysis and experience. The planner would have a different role. He or she would need to take a first cut at the required performance, understand and anticipate the decisions to be made, then work with designers on the trade-offs. The planner might start with the following:

- *What is the maximum anticipated production rate, in pounds per hour?*

- *Over some future period, what will be the range of products to be frozen?*

- *For each product, how rapidly must freezing take place—to maintain quality, minimize weight loss, and satisfy buyers' requirements?*

- *Are there reasons to consider cryogenic refrigeration for this job?*

- *How will freezing times vary with conditions that are still to be decided—packaging materials, product size, and freezer operating temperatures, among many others?*

- *Where is the freezer located in the plant, and how will the frozen product make its way quickly to a cold-storage warehouse?*

The resulting goals or targets would likely then set off some discussions. For example, the planner specifies a very fast, ideal freezing rate for his headed and gutted salmon line. The vendor comes back with a freezer that can actually do that. It is large, well-controlled, and expensive. The planner says, well, maybe we don't need to freeze quite that fast. And so on, through dozens of such trade-off decisions.

This chapter fills in a bit more information to allow the planner to better engage in discussions and decisions.

LOCATION

It's easier to specify the freezer location when the new plant is still on the drawing board. More common for small and existing plants is to find a spot within a structure that is already there. There are several things to think about—many are obvious.

Floor Space

One of the selling features of spiral freezers is the small footprint, when compared with an equivalent tunnel freezer. One cost of the smaller footprint is likely to be the ceiling height required by the stacked spirals. And as noted elsewhere in this manual, both production rate and freezing time will affect the size of the freezer, whatever type is selected. Future plans, too, must fall into floor space requirements, if a freezer expansion (or add-on) is to take advantage of a site already selected.

Foundation

There is an obvious need for a foundation strong enough to support the freezer, and that would be addressed by the design engineers. Under some circumstances, there is also a concern with moisture or groundwater under a freezer. When operated over some period of time, this moisture will freeze, causing cracking or distortion of the foundation. It is considered an industry standard to provide both insulation and a heat source in a properly designed freezer foundation (Cieloha 2005). Some further details and diagrams of these practices appear in *Planning Seafood Cold Storage* (Kolbe et al. 2006).

Proximity to Machinery

The freezer and refrigeration cold-sink must be connected but not necessarily located at the same spot. Smaller freezers will commonly have a **freon** unit directly attached, or close by. Or the evaporators (heat exchangers withdrawing the heat of freezing) will be directly attached and connected to a **condensing unit** (compressor, air-cooled condenser, reservoir, and controls) on the roof directly overhead.

For large ammonia systems, the evaporators within the freezer connect to the refrigeration machinery located in a separate engine room. The length of the piping connecting the two sites relates directly to costs—costs to performance due to heat leakage and pressure drops, costs of insulated piping materials and assembly, cost of refrigerant due to increased volume of piping and reservoir. (The relatively low cost of ammonia is one of its many advantages.)

If the refrigeration is driven by cryogenics, the distance between storage tanks and freezer is equally significant. Particularly for liquid nitrogen, heat gains in the supply lines represent a high loss of the –320°F refrigerant and, therefore, cost. Pipes must be very well insulated. For lines exceeding 100 feet or so, experiences of some designers indicate that special vacuum-insulated lines become cost-effective (Weiner 2003).

For any of the freezers, ready access to water and electric power can be a significant consideration.

Position in the Production Line

How the product flows efficiently from one process to the next is a favorite topic of industrial engineers. Although most existing plants don't have the luxury of optimal layout efficiency, having space available before and after the freezing station is critical. Certainly, space is needed in front of the freezer for staging, packaging, loading onto racks, or whatever. Aisle width is required for forklift trucks, and space is needed for storing folded boxes, etc. Product quality requires good temperature control for staged product waiting for the freezer.

After freezing, there are in particular two stations that can also dramatically affect product quality—these are glazing, if done, and cold storage.

Flow to the Cold Store

Proximity is the layout issue, but time and temperature are the important factors. Leaving the freezer, a product's core temperature will be somewhat warmer than its outer surface as well as the air, brine, or plates inside the freezer. Within just a few minutes, heat from the warm core flows to the cold surface, evening out the inside temperature to some average. At the same time, the product starts to warm up. Consider the block of mince previously tracked through the plate freezer in Figure 1-2. After 90 minutes of freezing, its core temperature was about 2°F, and its surface about –17°F.

Assume that same block is then removed from the plate freezer and set on a conveyor in a 60°F room for 30 minutes prior to pack up. Figure 4-1 simulates the results. After 10 minutes, temperatures within the block have already equilibrated and started to increase, as heat flows in from the warm room. Within 30 minutes the outer surface has increased to about 12°F.

A rapid temperature increase is not unexpected. As the next section shows, the amount of heat energy needed to change temperature in this frozen zone is relatively small. Additionally for this frozen material, the thermal conductivity k (a measure of how easily heat flows through) is quite large—about four times what it was in the unfrozen material. So whatever heat gets in will spread quickly throughout the product.

The rapid rise will be even greater in thin products. In one example, small, 3.5 to 4.5 ounce frozen fillets left a spiral freezer and dropped from its conveyor into a large tote positioned underneath. The measured core temperature of around –15°F increased to almost 25°F within 10 minutes (Kolbe and Cooper 1989).

In a second example of a poorly controlled process, pink shrimp were IQF frozen in a spiral freezer, sprayed for a glaze on exit, and then bagged in 1 pound lots. Farther down the conveyor, bags were packed into corrugated cartons and stacked on a pallet. When full, the pallet went into a cold storage room poorly controlled at about 14°F. Temperature monitors inside the bags showed the shrimp to be partially thawed at 23-25°F. After 7 days of storage, shrimp at the pallet's center continued to be above 20°F. Such a long time in the red zone would have supported ice crystal growth, enzymatic reactions, drip loss, texture and taste

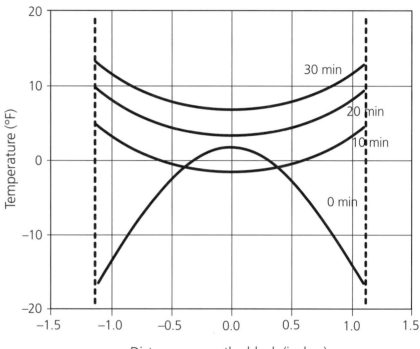

Distance across the block (inches)

problems—extremely poor quality. (This plant is no longer operating.)

In these cases, much of the freezer's good work is lost on the way to the cold store. Food technologists recommend that frozen product be kept close to the cold storage temperature as it enters. Although engineers typically design the cold store's refrigeration to remove about 10°F excess heat, too much will overload the system, cause temperature fluctuations, and in worst cases, very slow freezing rates of new product that has partially thawed. This highlights the need for processing room temperature control (less than 50°F, as recommended by Persson and Löndahl 1993) and rapid transport to the cold store or to some intermediate buffer storage.

The Glazing Step

Many unpackaged frozen seafood products are glazed at the exit end of the freezer. This results from a spray or bath of cold water, often mixed with a little sugar or salt to toughen the ice coating. The glaze creates both a moisture and oxygen seal while in frozen storage, thus supporting a good-quality product.

Unfortunately, it also provides heat, which tends to increase the product's temperature before heading to the cold store. Figure 4-2 shows core temperatures of frozen salmon after dipping in a glaze tank. A potential post-freeze step is to take the temperature back down after glazing. This seems to be often recommended but infrequently done.

Figure 4-2. Fish core temperature rise due to glazing. The freezing curves show H&G sockeye salmon (averaging 6½ lb) in a poorly balanced blast freezer. All are removed and glazed after around 11 hours, then boxed and placed in a –10°F cold room for the next 6 hours.

(from Kolbe and Cooper 1989)

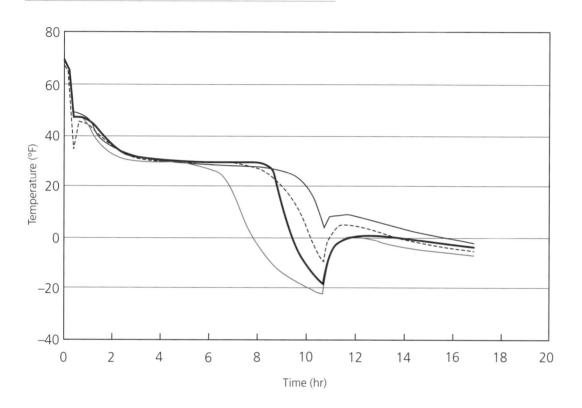

Large fish would have a small surface-to-volume ratio; temperature increase can be relatively small. But for small items like IQF shrimp, which take on a 5-10% glaze, temperature rise can be significant.

Löndahl and Goransson (1991) described the quality improvement of peeled frozen pink shrimp (*Pandalus borealis*) as the result of refreezing after the glaze step. In those tests, glazing raised the temperature of shrimp leaving the freezer from –4°F to 25°F; they were then packaged and stored at –13°F for a 5-week period. Without a step to refreeze after glaze, the shrimp taken from storage and then thawed had a 7% drip loss. A second line included a freezer to quickly reduce the 25°F glazed shrimp to 0°F before storage; when these shrimp were thawed and tested, drip loss was just 1%. Sensory panel tests also showed the refrozen product to be superior. The slow re-freezing that occurred in the cold store supported ice crystal growth and tissue damage. At the very least, this damage significantly reduced the final weight of a very high-priced product.

FREEZING CAPACITY

This term **freezing capacity** is confusing. On one hand is the freezer; on the other hand is the refrigeration machinery. Often the term refers to the number of pounds per hour you can push through the freezer. Converting these pounds per hour to a frozen product, added to some other heat loads removed in the freezer, creates the **refrigeration capacity**. This is the rate of heat absorbed by the refrigeration machinery. **Rate of transferring energy** is the engineering definition of **power** (see Appendix). So freezing capacity relates directly to the required size of the refrigeration machinery— i.e., how much horsepower is involved.

The contractor would specify a freezer that will likely do the job—that is, handle the seafood freezing rate required. He would also give the refrigeration power required—from the machinery already in the plant or from machinery that will have to be added. It can be valuable for the planner to know where these capacity numbers come from.

Figure 4-3 is a graphic display of the heat being removed at each temperature from a freezing seafood sample. The total area under the curve represents the amount of heat energy that has to be removed to freeze the product—a 1-pound package of surimi in this example—from 50°F to –15°F. As indicated by the shaded area at the right, one must remove about 21 Btu of heat energy to prechill this package to its initial freezing point of 29°F. Then, over a short temperature span from 29° to about 20°F, most of the water (about 75%) freezes to ice as an additional 96 Btu, mostly the latent heat of fusion, is removed. Only another 31 Btu must then be removed to take the now mostly frozen product down to its final average temperature, which we assume here to be –15°F. Persson and Löndahl (1993) report that "international definitions and standards" dictate that freezing be carried out to an equalization temperature of 0°F. The actual equalization temperature should roughly match the required product storage temperature that might fall in the range 0 to –25°F. It may depend on what the customer specifies. (Note: The curve in Figure 4-3 represents actual measured heat removed at each temperature and is used here for illustration. Engineers commonly use a simpler procedure to sum up

energy removed from the three chilling/freezing zones, Krack Corp. 1992.)

When added together, the total heat energy to be removed from the pound of product is 148 Btu. Although this figure was measured for a moisture content of 80%, it could apply as a conservative (i.e., giving adequately sized equipment) figure. In fact, as moisture content falls, the freezing energy will fall as well. For example, measured moisture content of albacore tuna ranges anywhere from 58 to 72%, depending on the lipid content (Morrissey 2005). Using calculation procedures of Krack (1992), a fish with a moisture content of 67% would have its freezing heat close to 128 Btu per pound.

To find the rate that the heat must be removed, multiply the unit of heat by the desired rate of production. For example, if a small tunnel freezer is to process 200 pounds per hour, the freezing capacity would be

$$\frac{200 \text{ lb}}{\text{hr}} \times \frac{148 \text{ Btu}}{\text{lb}} = \frac{29{,}600 \text{ Btu}}{\text{hr}}$$

This represents the absolute minimum required refrigeration capacity of this freezer. The rate is equivalent to 2.5 refrigeration tons (TR). As shown in the Appendix, 1 refrigeration ton = 12,000 Btu per hour. One refrigeration ton is the rate of heat removed when a ton (2,000 pounds) of ice melts in 24 hours.

It turns out that a refrigeration plant capable of removing heat energy at the rate of 2.5 TR will not keep up. This is because there are additional heat loads on the system, and the refrigeration machinery must remove those as well.

In a plate freezing operation, the "additional loads" include the heat that is in the structure (relatively warm plates, frames, housing), heat leakage from the surrounding room, and heat added by pumps moving the refrigerant. In fact, for plate freezers, these sources could contribute anywhere from 5 to 20% of the total load (Graham 1984, Cleland and Valentas 1997).

Brine freezers might be similar, with perhaps 10% of the heat leaking in from the surroundings, 5-15% added by the pumps, with freezing fish supplying the balance.

Blast freezers use powerful fans to circulate air. Because all the energy that goes into those

Figure 4-3.   Heat capacity of 1 pound of surimi with 80% moisture content; 8% cryoprotectants.

(from Wang and Kolbe 1991)

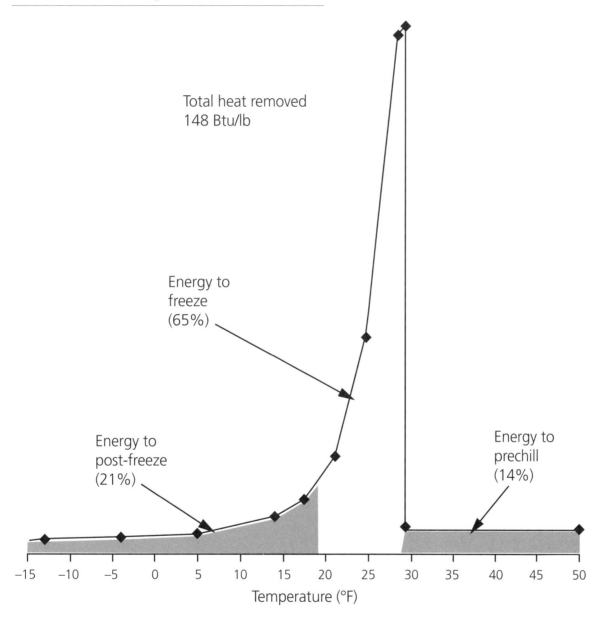

Total heat removed
148 Btu/lb

Energy to
freeze
(65%)

Energy to
post-freeze
(21%)

Energy to
prechill
(14%)

Temperature (°F)

fans ends up as heat in the freezer, fans represent additional loads that can be 30-40% of the total refrigeration requirement. If heat loads of a representative tunnel freezer are

Fan load	30%
Heat leakage from outside	10%
Defrost	5%
Fish	55%

then the total refrigeration power required in the previous example would be

$$\frac{2.5 \text{ TR}}{0.55} = 4.5 \text{ TR}$$

One rule of thumb used for plate freezer systems is that 35 TR is required to freeze 1 ton per hour of fish. For this blast freezer example, which includes the high fan-heat load, the figure becomes closer to 45 TR per ton per hour.

So the system design must balance freezing capacity with the refrigeration capacity. The latter can also be somewhat controlled, but with trade-offs. For example, to speed things up, the production manager might ask the operating engineer to crank down the temperature of the (low side) refrigerant flowing through the freezer. But this would come at a cost to the available refrigeration capacity. Recall Figure 3-3 showing how refrigeration capacity takes a hit as temperatures fall. The curve shows that decreasing a –30°F operating temperature by 9°F will diminish refrigeration capacity by 17%. The high side condensing temperature, too, can influence capacity, although to a lesser extent. For example, if the rooftop condenser on a hot day is allowed to increase by just 2°F, refrigeration capacity can fall by about 1% (Cleland and Valentas 1997).

POWER REQUIREMENTS

The freezing operation will likely be the biggest electrical power user in the plant. The actual power use will vary with the production schedule. So predicting can be difficult. For a batch freezer such as an air-blast cell, power (i.e., the rate of energy consumed) will be high at the beginning of the freezing cycle and taper off as the rate of heat removed tapers off. But often separate freezers are phased in at different schedules, each with high initial refrigeration loads as well. For ammonia systems, different combinations of screw and reciprocating

compressors are switched on and off as load goes up and down. Individual compressor efficiencies can vary with load, especially with screw compressors. Defrost schedules may change with products and freezing practices. Overhead loads will add to freezing power in a variety of forms—lights, building heat, battery chargers for forklift trucks, and others.

Amid that confusion, however, it is possible to make some estimates—i.e., some comparative average power requirements assuming continuous freezing, as on spirals or belts. One of the major variables is the refrigeration system's coefficient of performance (COP)

$$\text{COP} = \frac{\text{What You Get}}{\text{What You Pay For}}$$

"What you get" is the refrigeration tons removed in the freezer—the capacity that freezes the fish and absorbs other loads like fan heat, leakage, and defrost. "What you pay for" is the power required to drive the compressors. To calculate COP, both of these quantities are put in the same units of "energy transferred per time" (see Appendix). Typical COP values for freezing might be in the range of 1-2. This depends a little on the refrigerant used, and a lot on the compressor efficiencies, evaporator refrigerant temperature, the lift between the evaporator and condenser, among other factors.

Cleland and Valentas (1997) give performance information for a "good practice" two-stage refrigeration system, and this was used to create Figure 4-4. The temperatures shown are for the evaporating (suction) and condensing (discharge) refrigerant. The actual freezer temperature (e.g., air-blast or plates) might be around 5-10°F warmer than the suction temperature. For an air-cooled condenser, the air temperature would have to be about 10-20°F cooler than the refrigerant temperature shown.

In Table 4-1, we use these estimates to get some ballpark values for refrigeration power consumption. The requirement in this example is to freeze 1 ton per hour of fish having a moisture content of around 80%. For a continuous freezer, as a tunnel or spiral, the product is assumed to flow in and out continuously. For batch freezers, such as a horizontal plate freezer or blast cell, an average freezing rate is calculated as follows.

$$\frac{\text{Average}}{\substack{\text{Freezing} \\ \text{Rate}}} = \frac{\text{Tons of Fish per Batch}}{\text{Freezing or Cycle Time, in Hours}}$$

Refrigeration compressors will be driven by electric motors. Total electric power costs will depend on the system COP, as described in the example two-stage system of Figure 4-4. Notes explaining assumptions and calculations follow the table.

Actual energy costs will be greater than those shown in the above table. Such overhead loads as lights, heat, lift truck chargers, conveyors, compressed air controllers, and many others must be included in any energy assessment. And many operations often run inefficiently—partially filled blast freezers, a lightly loaded screw compressor, shut-down of freezer delayed to coincide with work shifts, and others. The intent of Table 4-1 is to compare systems rather than present absolute values.

Figure 4-4. Coefficient of Performance estimates for a two-stage refrigeration system.

(from projections of Cleland and Valentas 1997)

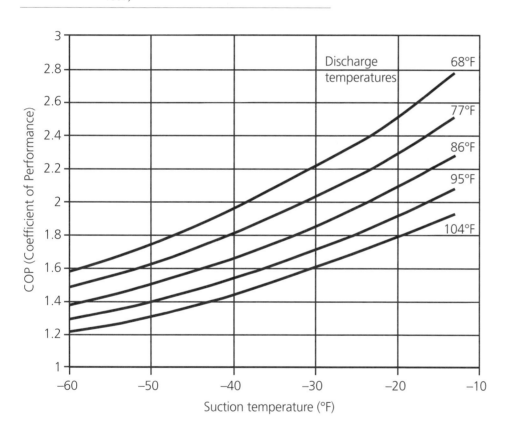

Table 4-1. Estimated power requirements for freezing operations

Assume freezing capacity of 1 ton of fish per hour.

	Coefficient of performance (COP)	Required capacity in refrigeration tons (TR)	Horsepower required to drive compressors (HP)	Kilowatts of power driving compressor motors (kW)	Kilowatts of power driving additional loads (kW)	Total kilowatts of power (kW)	Electrical energy required per pound frozen (kWh/lb)
	2	37	87.3	72.4	18.1	90.5	0.045
Brine	2.2	37	79.4	65.8	16.4	82.2	0.041
freezer	2.4	37	72.8	60.3	15.1	75.4	0.038
	2.6	37	67.2	55.7	13.9	69.6	0.035
	1.4	35	118.0	97.8	14.7	112.5	0.056
Plate	1.6	35	103.2	85.6	12.8	98.4	0.049
freezer	1.8	35	91.8	76.1	11.4	87.5	0.044
	2	35	82.6	68.5	10.3	78.7	0.039
	2.2	35	75.1	62.2	9.3	71.6	0.036
	1.4	45	151.7	125.8	71.7	197.4	0.099
Blast	1.6	45	132.7	110.0	69.3	179.3	0.090
freezer	1.8	45	118.0	97.8	67.5	165.3	0.083
	2	45	106.2	88.0	66.0	154.0	0.077
	2.2	45	96.5	80.0	64.8	144.8	0.072
	Note 1	Note 2	Note 3	Note 4	Note 5		

Note 1: Brine freezers must operate at a relatively high refrigerant temperature to prevent freezing the chiller. Sodium chloride brine has a minimum freezing temperature of –6°F; warmer as it becomes diluted. Thus the refrigerant temperature will be on the order of –10°F or warmer. The machinery is relatively efficient at these temperatures, so the range of expected COP values will be greater. The capacity of the machinery, too, will be greater, as shown in the previous chapter (Figure 3-3).

For blast and plate freezers, evaporating refrigerant temperature will be controlled to a far lower value—in the range –30 to –40°F. Expected values of COP, and the refrigeration capacity of the machinery, will be lower.

Note 2: The previous section showed how the required refrigeration capacity must remove the heat not only from freezing fish, but from a number of other heat loads as well. A capacity of 35 TR might be considered a decent rule of thumb for a 1-ton-per-hour production rate in a plate freezer. This might tend to be similar to that of a brine freezer. Its higher operating temperature would cause the heat leakage load to be lower, but brine circulation pumps would add additional pump heat. A refrigerating load of 37 TR is assumed.

For blast freezers, a value of 45 TR for a 1-ton-per-hour rate was calculated in the previous section. It is higher due to the added fan heat load.

Note 3: COP is defined as $\dfrac{\text{Refrigeration Capacity}}{\text{Power Driving Compressors}}$ and uses the conversion, 1 TR = 4.72 HP (see Appendix).

Note 4: The efficiency, E, of large motors is on the order of 90% and is defined as $E = \dfrac{\text{Power Supplied by the Motor}}{\text{Power Supplied to the Motor}}$

This step uses the conversion 1 HP = 0.746 kW (see Appendix).

Note 5: Most of the electrical power will drive the refrigeration compressors. However, all refrigeration systems will require additional electrical power to drive various other components. Depending on the system, such loads as condenser coolant pumps or fans, pumps for liquid refrigerant, oil and other loads represent 10-15% of the power driving compressor motors (Cleland and Valentas 1997). We'll use 15% for these calculations.

For brine freezers, an additional load would be brine-circulating pumps. Assume these to be another 10% of the compressor load.

Air-blast freezers have an additional load, which will be significant. This is the load on motors driving the circulating fans; for spiral or tunnel freezers, the conveyor drive motors will consume some more. A value of 30% of the total load coincides with the calculations in the previous section and contributes to the figures in this table.

ENERGY CONSERVATION

Freezing will probably be the largest electrical energy user in the plant. But the cost of electricity remains a small piece of the total processing costs when compared to fish, labor, shipping, packaging, taxes and fees, and so on. Energy conservation doesn't get a lot of attention, particularly in small plants in the Pacific Northwest where electric rates are relatively low. But there are reasons to look seriously at energy conservation.

- *Energy costs will continue to climb; in some areas such as Western Alaska, energy already is a major expense.*

- *It is an expense that the plant operator **can** do something about, often with a small investment and short payback period.*

- *Energy saved at one operation can often be used in another. The result would be lower energy costs while sometimes avoiding the cost of installing new equipment.*

- *It is environmentally responsible.*

Several things can be done both with freezing technology and with the refrigeration equipment to lower the consumption and cost of energy (Wilcox 1999, Gameiro 2002). Some measures require a major investment, with the payoff period depending strongly on the local unit cost of electrical energy. Other measures involve simpler alterations of the process, creating energy payoffs at relatively minor cost.

The following measures, applying primarily to vapor-compression refrigeration systems, will get results.

Freezer Design and Operation

Earlier sections noted that a good share of the freezer's heat load results from sources other than freezing products. These could be fans, defrosting, or leakage of air or heat into the freezer compartment. This suggests several measures that will reduce energy consumption.

- *Reduce freezing time. Anything that reduces time in the freezer will also reduce the contribution of extra heat loads. Modifications that can lower freeze times have been covered in previous sections.*

- *Reduce the rate of heat leakage. Erecting enclosures (for example, around a plate freezer), increasing*

insulation thickness, and redesigning doors to allow quicker loading and less air infiltration, will all improve energy efficiency.

- *Select an oversized evaporator. This applies primarily to blast freezers. The larger the heat transfer area in the evaporator, the smaller the temperature difference between the freezing medium (such as air) and refrigerant. This enables several possibilities—a lower temperature leading to a faster freeze, a lower air velocity (and thus fan energy), and operation at a higher refrigerant temperature. The last measure increases the refrigeration capacity (Figure 3-3) while decreasing the power (kW) needed to remove a ton of refrigeration.*

- *Modify the defrost schedule. Defrosting adds heat to the system. Often, defrosting occurs on a regular basis; for example, after each freeze cycle. And in some cases, this could be modified—stretched out—for greater energy efficiency. Measure the freeze-time effects of delaying defrost, then adjust the manual or automatic control schedule accordingly.*

- *Control speed of evaporator fans. The power required to drive fans will vary with (speed)[3]. It is this power that becomes heat that must be removed by refrigeration. This means that if fan speed is adjusted to half, airflow will fall to about half as well. But power driving the fans (and ending up as heat) will fall to one-eighth. Fan heat in blast freezers can be decreased with the use of fan speed control; more detail appears in a case study, below.*

Refrigeration Machinery Options

Today's refrigeration engineers are particularly conscious of maximizing energy efficiency. For existing systems that may have developed piecemeal over the years, the list of changes/retrofits can cover a vast range. For new systems, there are some basic energy conservation design measures that frequently make the list. Involvement of refrigeration and energy efficiency engineers will fill in the details.

- *Maximize the refrigerant evaporating temperature. Each 2°F increase of refrigerant temperature in the evaporator can decrease energy consumption by roughly 2-3% (Stoecker 1998). Look for such opportunities in systems that support multiple-temperature freezers, and in freezers that could operate at slightly warmer temperatures.*

- *Minimize the refrigerant condensing temperature. Anything that can reduce condensing temperature can reduce energy consumed, 2-3% for each 2°F reduction (Stoecker 1998, Gameiro 2002).*

- *Control condenser size and fan speed. Ammonia condensers are typically evaporative air-cooled units; an oversized condenser will remove the necessary heat load using less horsepower from fan and pump motors. Capacity-control using variable speed drives on the fan motors is also important, as fan power falls with the cube of speed. And choose axial vs. centrifugal fans for greater efficiency (Wilcox 1999).*

- *Use automatic gas purgers. Non-condensable gases in ammonia refrigeration systems will hurt performance.*

- *Diversify compressor selection. Capacity control typically used on screw compressors to unload capacity makes them very inefficient in the low-capacity range. However, screw compressors are **very** efficient at full speed and capacity. Energy efficiency will result from an engine room design that uses a larger number of smaller compressors, one or two of which have capacity controlled by variable speed motors.*

- *Employ automatic controls. Experience of designers has shown significant savings with centralized computer control of the system. Programs can sense changing loads and ambient conditions to continually optimize the operation.*

- *Use adequate piping. Both size and insulation can be important. Long runs of piping can allow heat leakage and pressure drops that will cost efficiency points. In one example, allowing pressure drop in the suction line to increase from 1 to 2 psi increased energy use by 5% (Gameiro 2002). Allowing an excessive increase in refrigerant temperature leaving the evaporator (i.e., superheat) will cost energy as well—about 0.4% for each 2°F.*

- *Recover waste heat. The largest source is condenser heat that can be used to complement energy needs elsewhere in the plant—for room heating, foundations under cold storage rooms, preheating boiler supply water, cleanup water, and others.*

A Blast Freezer Project

A recent project with blast freezers demonstrated significant savings that could result from two actions: one was to control the flow and balance of air through the racks of product; the other was to apply speed control devices on fans (Kolbe et al. 2004a).

A stationary blast freezer processing 22-pound cartons of sardines in 19,000-pound lots was first modified to improve efficiency and to conserve energy. The addition of strategic baffling produced a uniform and slightly increased airflow. Maximum measured freeze times of 12.5 hours fell to 10.5 hours; total electrical energy savings was estimated to be 12%.

The second action was to then install a variable frequency drive (VFD) to slow evaporator fans during the freeze cycle. In one example run when fans were slowed to 75% speed after an initial freezing period, maximum freezing times increased back up to about 11.5 hours, but the overall energy consumed fell by an additional 10%. Math models developed from these experiments indicate a more optimum solution, in which the measured 10% energy savings increase could be attained with just a 2% increase in freezing time.

PLANNING FOR ONBOARD SYSTEMS FOR FISHING BOATS

Most of the planning concepts already addressed in this manual would cover systems aboard fisher-processor vessels. But there are some additional issues for small boats in the range of 40-60 feet. These are often involved in freezing salmon, albacore tuna, prawns, and other species, landed and processed at a rate of a ton or two per day. For these small processors, there may be some additional questions to ask.

The first might be, why freeze at sea? Among the reasons is quality. When frozen correctly, the final product can be as good as that freshly caught and certainly better than an iced product held several days in the hold. Decreasing the time of unfrozen storage at sea also extends the shelf life—the time available for distribution of the thawed product in the shoreside marketing system.

A second reason is price. A well-treated frozen product ready for shipment can bring a better price. And keeping a frozen product on board may enable the fisherman to hold the product until some later time when the price is up.

A third reason is flexibility. For example, long holding times make it possible to fish until the hold is full, thus reducing costs of both fuel and time spent for extra trips home.

The type of freezers found on small boats are among those covered earlier— air-blast freezing of fish that are hung or laid out on racks or shelves, contact freezing with fish on freezer-plate shelving, and sodium chloride brine freezing with fish in an immersion bath or spray.

Which to use, and how to plan? The best resource for learning the important issues of at-sea freezing is a fisherman who is currently doing it. A few points stand out as worthy of some attention:

- *What freezing technique is suitable for the various species you wish to catch—this season and next? For example, if considering a brine freezer, what about salt penetration in the fish—particularly those that have been gutted? Is bagging required or an option? What would be the expected life of the subsequently stored frozen product? If blast freezing is used, what about dehydration?*

- *In what form will the potential buyer accept frozen product? Some buyers won't accept boat-frozen salmon unless they really know the fisherman and know that the product has been well cared for. In another example, certain Japanese blackcod markets have specified block-frozen dressed fish; a brine-frozen product would not sell.*

- *What additional onboard handling (labor) will be necessary to prepare, freeze, and store the product?*

- *Is prechilling needed before the product is frozen? How long must fish lie on deck before freezing? Boat-frozen fish can be of superior quality to those landed in ice, but they need to be fresh and well cared for prior to freezing. Albacore tuna, which can arrive on deck with a core temperature near 90°F, requires prechilling. Prechilling also helps to retard the onset, and decrease the effects of, rigor mortis. If prechilling is required, will ice or refrigerated seawater (RSW) be workable on board?*

- *What daily catch rate does one use for sizing the freezing system? If the catch rate begins to exceed that figure, are you willing to stop fishing? If not, what might be the effect of an overloaded system on quality, and how will that affect customer acceptance?*

- *Are good refrigeration contractors available who can design, install, and fix successful systems? Will these contractors or other service people be around for*

parts and assistance in case of a breakdown during the height of the season?

- *Have all the costs been considered—equipment purchase and installation, extra power supply (perhaps an auxiliary-driven hydraulic pump or generator for operation while the main engine is shut down), conversion and insulation of the fish hold, extra fuel costs (vs. savings on ice), future maintenance costs, extra crew needed for preprocessing and handling?*

Freezing systems presently available for small boats consist either of several different components scattered about the boat (each component sized to work at optimum capacity) or a series of prepackaged units requiring a lesser amount of installation effort on the boat.

Blast Freezers

There is lots of experience on the U.S. and Canadian West Coast with small-boat blast freezers on salmon trollers. Each species and product form may have different optimum holding and handling conditions, topics covered in Chapter 2. Some recommendations for blast freezers are covered in Chapter 3. For installations in a fish hold, the small space creates some difficulties, and a few recommendations should be highlighted.

- *The air blowing over the fish spread out on shelves or racks should not be warmer than –20°F. According to experience in Canada and Europe, velocities should be greater than 400 feet per minute, a bare minimum. Good blast freezer design calls for a velocity of 1,000 feet per minute or better. Slow-moving air makes freezing times too long (although too high a velocity means too much power is being supplied to the fans, and this means too much added heat.)*

- *All shelves must get the same flow rate of air. Freezing rates will vary quite a lot with air velocity, so proper fan size, ducting, and adequate spacing between shelves are all important.*

- *The core of the fish (the warmest part) ought to be down to –5°F before moving the fish to a separate storage area. Freezing can take a while. For example, Canadian experience with 10- to 12-pound salmon (thickness of about 3.5 inches), blasted with –30°F air in a good freezer will reach the required core temperature in about 5 hours; they should not be removed from the freezer until then.*

- *When defrosting by hot gas or electricity, see that the fans switch off to minimize warming in the hold.*

- *Store product in an area separate from the freezer, maintained if possible at or below –20°F. Although difficult in a small boat, the ideal is to maintain this area at a steady temperature (fluctuation less than about 5°F). The manual Planning Seafood Cold Storage (Kolbe et al. 2006) covers quality loss in frozen storage.*

- *Minimize dehydration during storage. In many cases, this is by glazing—dipping the frozen fish in a cold-water bath that sometimes has a little sugar thrown in to make the glaze less brittle. Packaging the product in sealed plastic bags of appropriate material might be another option.*

- *The fish hold must be well insulated—4 inches or more of urethane foam is minimum; 6 inches on the engine room bulkhead. See that storage racks are built so that frozen fish don't lie directly against the fish hold walls. Heat leaking from the outside would warm up those fish that aren't well exposed to the cold air in the room.*

Blast freezer systems can be pieced together from separate components, or available in the form of "packages" advertised as being commonly available for marine use. This implies special factors that enable longer life under the corrosive environment and hard-use conditions existing on a boat—such as special coatings, avoidance of dissimilar metal contacts, use of sealed motors, and rugged controls. One example of a package is the hatch-cover unit of Figure 4-5. Self-contained, it can be installed when needed and operates when power and condenser water are connected. Additional requirements will include a power source (electric generator, hydraulic pump, diesel engine), racking and storage structure, hold insulation, and possibly additional on-deck refrigeration.

Contact Plate Shelf Freezers

In small-boat systems, shelf freezers refer to shelves made of hollow freezer plates. These plates are the evaporator of Figure 3-2. The refrigerant boils away as it flows through passages within (Figure 3-9).

Figure 4-6 shows a small cabinet shelf freezer capable of freezing 1,000-2,000 pounds per day. While the freezer is not equipped with fans, the cold air falling from the finned surface above serves to hasten freezing.

Figure 4-5. Hatch-cover package blast freezer. After mounting, the unit would require only power and water hookup to operate.

(courtesy Integrated Marine Systems)

750 lb/day System

Brine Freezers

Freezing in a sodium chloride brine bath has the advantage of speed—with cold liquid all around the fish, heat transfer rates are very high. And with brine you can actually get more out of your refrigeration machinery—the warmer operating temperature of sodium chloride brine freezers means higher suction temperature and pressure and, therefore, a higher capacity.

In the Pacific Northwest, brine is most commonly used onboard small trollers landing albacore tuna, a product that is frozen round. Salt uptake is a minor factor if the landed product will be canned—the skin and outer layer are discarded. Other species such as blackcod have also been successfully brine-frozen aboard the vessel. Packaged units are available for brine freezers as well. Figure 4-7 shows such a system with compressor, condenser, reservoirs, controls, and brine chiller all mounted on the same frame.

Onboard brine freezing systems using immersion tanks frequently have essentially two refrigeration circuits (or even two separate systems). One goes to the tank, the other to a fish hold unit (ceiling coils, plates, or a blower unit), which maintains

Figure 4-6. Shelf-freezer cabinet. The enclosure on the left houses the motor-driven compressor and controls.

(courtesy Gunthela Enterprise, Ltd.)

a low storage temperature approaching –20°F. Most companies and contractors that install blast freeze systems can also supply the controls and equipment needed to include a brine freezing tank.

Some brine freezers onboard albacore boats do not use a tank at all. Instead, the brine is sprayed over the product stacked in the hold. After trickling through the fish, it collects in a sump in the fish-hold floor. From there it is pumped through a heat exchanger, then back through the sprayers, and the cycle continues. In this case, the storage temperature can be no lower than around 0°F, the practical lower limit for sodium chloride brine.

For some fish, like salmon, brine freezing without packaging is not an option. For albacore tuna,

it can be effective. Some comparisons can be made of the two systems—blast and brine—when both are an option.

Air-Blast Benefits

- *An air-blast freezer can be designed for one fishery (e.g., albacore), but is an option for others.*

- *Frozen products can be very high in quality and attractive to a greater market range.*

- *One has access to the hold while freezing is under way.*

- *Addition of a good glaze is possible after freezing.*

- *System temperatures can be adjusted.*

Figure 4-7. Brine chiller package. Electric motor-driven compressor is mounted next to the refrigerant reservoir (receiver). Two barrel chillers are mounted in parallel.

(courtesy Integrated Marine Systems, Inc.)

Air-Blast Costs

• *The refrigeration capacity of a given compressor is lower at the lower evaporator temperature.*

• *More product handling, thus more labor, is required.*

• *Dehydration of the stored product can be a problem.*

• *Periodic defrosting is necessary.*

• *Lower freezing temperature means a lower energy efficiency.*

Brine Benefits

• *High refrigeration capacity at high evaporator temperatures.*

• *Less requirement for handling, particularly with spray brine.*

• *No defrost necessary.*

• *System could be adopted for refrigerated seawater (RSW) chilling.*

• *Better energy efficiency.*

Brine Costs

• *Products such as albacore are suitable only for the canning market; salt uptake is problematic.*

• *Bacteria buildup and cleaning difficulties.*

• *Spray brine presents problems with uneven freezing; there is a need for antifoaming agents.*

• *Storage in –20°F air will cause dehydration of the product unless it is protected. Glazing of fish frozen in brine is said to be difficult because the glaze won't stick to the fish.*

• *Equipment corrosion*

• *Need to avoid copper alloys in the system to prevent discoloration of canned products.*

• *Need to buy and mix salt; must control concentration to avoid freeze-up in the chillers.*

• *Typical brine temperatures can be 5°F or warmer.*

• *Sprayers operating in the confined space of a small-boat fish hold cannot do a very good job of evenly distributing the spray. And plugging up with debris and cleaning/sanitizing are ongoing difficulties.*

Some Costs and Cautions

Here is a rough example of what onboard freezing might cost. These are ballpark figures for 2005.

Equipment

A new motor-driven blast-freeze unit capable of freezing around 5,000 pounds of fish per day would cost about $40,000. A typical configuration has the evaporator mounted in the fish hold; the condenser unit mounted in the engine room. Installation cost may add $10,000-$15,000, although any work done by the boat owner would decrease this figure.

The equipment cost of a hatch-mounted blast freezer would be greater (one vendor lists $47,000). But with only water and a power cord to attach, installation cost is relatively minor.

Other equipment items must be in place. One is power. The blast unit may have to have a generator that can put out 40 kW to get it started (with power falling to about 20 kW once freezing is underway). Alternatives could be hydraulics or diesel.

A second is the fish hold structure to support hanging or shelved fish in the uniform high-velocity air stream. Design depends on space available, freezing times, catch rates, and markets.

Fish Hold

The fish hold must be insulated, lined, and prepared for fish freezing and storage. In the Pacific Northwest, insulation is typically urethane, foamed in place, then shaved and shaped to a thickness of 4-6 inches. Costs are on the order of $1.30 per square foot per inch of thickness. The liner represents an even greater cost. A sprayed two-part polyurethane coating might run on the order of $10-$12 per square foot. A tougher, stronger fiberglass liner could be $15 or more. Installers report that material prices are increasing rapidly with the price of petroleum.

Cautions

An onboard mechanical refrigeration system can be expensive. One can make some shortcuts to bring these costs down, and having some of the work done by the boat owner will help as well. But there is a danger of cutting corners to an extent that you end up with an inadequate or marginally reliable freezing system. Work with experienced professional contractors, both for the insulation job and for the refrigeration system design and installation.

To re-emphasize a few specific cautions:

1. The capacity and temperature must be right. They depend at least on species, market, and catch rate. A used compressor, fan, or coil unit of uncertain capacity, a poorly designed homemade brine chiller, inadequate sizes of piping and valves, can all contribute to a system that just won't make it. The result is a landed product with a bad reputation.

2. Operation in remote areas requires maximum reliability. And this comes with use of well-designed, top quality equipment.

3. Controls are important. The decision on manual vs. automatic operation is up to the designer. But controls used must be reliable—high quality and of a rugged type and design that will be unaffected by the hostile environment at sea.

4. Oil return can be a problem. Refrigerant gas tends to carry along lubrication oil from the compressor. If the pipe sizing and layout aren't designed to continually return this oil, the compressor will be in trouble.

5. A dual (two-temperature) system can be tricky. Often a single refrigeration unit is called upon to do two things: one to freeze fish in one piping circuit where loads and temperature vary a lot; the other to maintain a fish hold storage temperature that should not vary much at all. Canadian investigators have advocated the use of two separate refrigeration systems to avoid these problems.

6. Refrigeration leaks can be a problem, not only because they might result in reduced (or zero) capacity at a critical time, but also because refrigerant gas can be a hazard to health. Halocarbons (freons) in their common form are not toxic. But the leaking gas is heavier than air and presents some risk of filling the sleeping quarters or fish hold on a small boat, thus displacing oxygen and threatening asphyxiation. If this leaking refrigerant gas finds its way into a diesel engine intake, it can cause accelerated wear of the engine; if burned in the cylinders, it will form phosgene, a poisonous gas.

7. It's very important that fish hold insulation is installed right. In a wooden boat, previously moist wood must be completely dry before it is covered up. Four to five inches of foam, usually urethane, is the recommended minimum for freezer boats in the Northwest. More is required in the warmer waters of the southern and Gulf coasts. Foam installed on a steel vessel must be thick enough to cover structural members. Bridging of the insulating layer by frames will provide a heat-conducting path that will substantially increase leakage in the hold and create local "hot spots" along the walls of the storage area.

Figure 4-8 shows the resistance measured for various fish hold wall test-sections. Diagram A shows the R-value for no insulation—a plywood liner fastens directly to the steel frames. Diagram B shows a little improvement when the air space is filled with urethane foam. Heat still conducts freely along the steel frames. This would be far worse if a steel liner were used in place of plywood or other surface. The R-value jumps dramatically in diagram C when the foam buries the steel frame by just an inch or so. Figure 4-9 predicts resistance of a larger fish hold wall section as thickness of urethane foam increases.

Figure 4-8. Fish hold wall sections. The numbers represent the resistance ("R-value") to the heat flowing in from the warm surroundings.

(from Wang and Kolbe 1989)

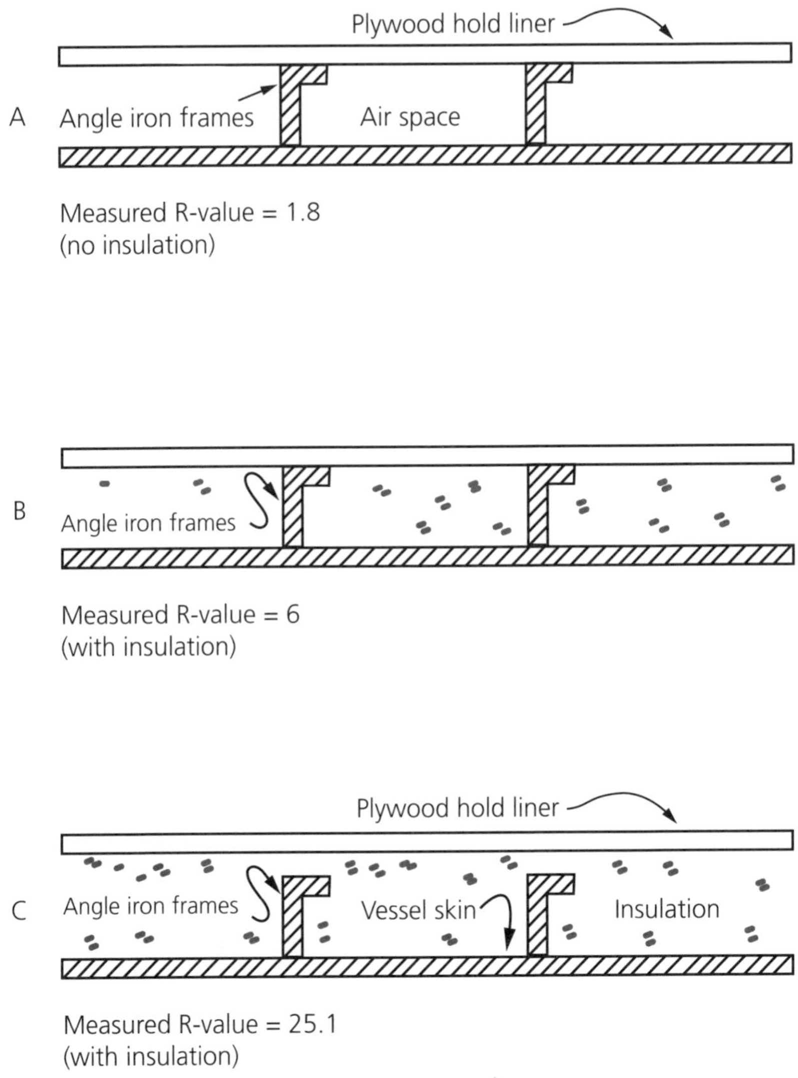

A   Plywood hold liner

Angle iron frames   Air space

Measured R-value = 1.8
(no insulation)

B

Angle iron frames

Measured R-value = 6
(with insulation)

C   Plywood hold liner

Angle iron frames   Vessel skin   Insulation

Measured R-value = 25.1
(with insulation)

Figure 4-9. R-value of fish hold wall sections. Insulation thickness, the liner material, and the way the liner is fastened to the vessel frame all contribute to the effectiveness of a fish hold's insulation system. At 4 inches, the insulation thickness equals the frame depth in this example. As the thickness increases, there is a dramatic jump in the R-value (resistance to heat flow). Note that the liner material is most important when the insulation layer is less than the frame depth.

(from Wang and Kolbe 1989)

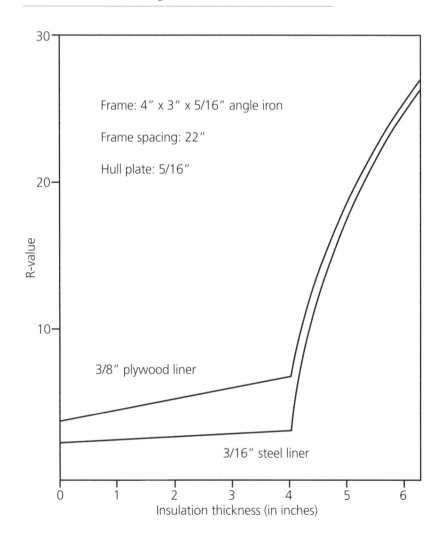

❄❄❄❄❄❄❄❄❄

Chapter 5   Scenarios

Freezing equipment appropriate for four different scenarios appear in this chapter. These situations are intended to be a representative plan, targeting smaller processors who are about to put in a new line.

Most of the costs and installation details came from contractors experienced with sales and design. Exercise caution when using costs; they reflect 2005 economics and are considered rough estimates. In most cases, used equipment could do the job at a cheaper cost. Success depends on what shape it's in. Used equipment may come with a critical need for local maintenance and repair services.

In addition to discussing freezers, scenarios discuss refrigeration needs, but details are omitted. In some cases, refrigeration will be purchased with the freezer; in others it comes as an addition to existing machinery.

BLAST FREEZER FOR HEADED AND GUTTED SALMON

Background

Leaders of an Alaska coastal village have generated funds to purchase and install a freezer. Its primary purpose will be to freeze headed and gutted sockeye salmon caught locally using gillnets. Currently, these fish are sold in the round to a large processor operating floating tenders in the bay. After chilling the catch in RSW (refrigerated seawater) tanks, the tenders periodically deliver to a large processing plant in another community.

A new plan would call for freezing at least part of the daily catch with the expectation of then selling the fish for a higher profit margin. With construction of a local cold storage warehouse in the future plan, high-quality frozen fish can then

support a village workforce producing other value-added product forms.

A target is to process 40,000 pounds of round fish per day, a figure that represents about 12% of the local landings at the peak of a 12-week season. The larger whole-fish size is on the order of 7 pounds. After removing heads and viscera, the freezing load becomes 30,000 pounds per day (Crapo et al. 2004).

For those fish to be landed and frozen, processors plan to require fish boats to have ice, RSW, or CSW (ice-chilled seawater) systems onboard, and be required to chill the fish to a temperature below 50°F upon off-loading. Because of requirements for a high-quality product and because of the uncertain length of frozen storage, the freezing fish will be expected to reach a temperature of –20°F before storage at that same temperature.

Existing facilities include a processing plant, but no refrigeration machinery is in place. Once freezing and storage facilities are in operation, the plant manager anticipates other process opportunities with coho and chinook salmon, halibut, flatfish, rockfish, and blackcod.

The Plan

With expected freezing times of 4-6 hours, and with limited plant dimensions (ceiling and length of the room), tunnel and spiral freezer configurations were rejected in favor of stationary blast cells. Such cells also offer the future option of loading larger, slower-freezing fish, something difficult with tunnels and spirals. Off-loaded fish will travel first through a heading/gutting machine, then will be manually stacked on racks or trolleys that roll into the blast cells.

Following a freeze cycle, fish will run through a glaze tank and then into boxes and then to the cold room. Storage at –20°F will initially be in freezer vans that are periodically barged out. Plans for the completed project call for the near-future construction of an on-site cold storage warehouse.

Calculations followed by experiments show that 7-8 pounds of salmon, after heading and gutting, will blast-freeze on racks within 5-6 hours. Three blast cells will each be capable of holding 3,500 pounds of headed and gutted sockeye. At peak production, three shifts could freeze the 30,000-pound target capacity. (Alternatively, larger cells holding 7,000-8,000 pounds would allow freezing during two work shifts, leaving one cell to be defrosted and cleaned.)

Refrigeration machinery will have a 35 TR capacity, driven by one 100 HP compressor. This ammonia system, housed in a separate engine room, assumes adequate electric power is available. Other equipment will include blast freezer coils and fans, an evaporative condenser, pressure vessels, piping, etc.

The Major Pieces

Construction of blast cells
Three cells, two doors in each $50,000

Refrigeration machinery
FOB points of manufacture $195,000

Insulation
Pipe and vessel insulation allowance $38,000

Freight $24,000

Installation $100,000

Machine room items
Ventilation, ammonia detector $13,000

Total $420,000

Note 1: This assumes a single 100 HP compressor. The use of two 50 HP compressors would add reliability and may increase efficiency. It would also increase cost.

The contractor supplying these figures notes many components that are not included. Among them are

- *Wheeled racks for the blast cells.*

- *Any special adjustments to foundation or buildings; structural supports for machinery.*

- *Electrical work, such as temporary construction power, power wiring, some starters, control relays, and wiring.*

- *Water source for machine room.*

- *Taxes, permits, fees.*

Note 2: The installed cost of this ammonia-refrigerated blast freezer system is roughly equivalent to an installed halocarbon-refrigerated blast-continuous system estimated by another company. Their self-contained unit can be located outside the plant and requires only electric power and a foundation to support its 15-ton weight.

Discussion

A freezer project of this magnitude, starting from scratch, will have to include other phases, both in planning and budgeting. Besides offloading, processing, and worker-support infrastructure, two of these relate directly to the refrigeration capacity: a cold storage warehouse and an ice plant. Given adequate funding, these future phases would be most cost-effective if everything were constructed at once. Failing that, a second, lesser-cost option is to design the engine room and ammonia system to accommodate future refrigeration needs. Examples include an oversized condenser, tanks, piping, and headers.

However, if such an over-design was not possible for phase one, the contractor has suggested a possible plan, which uses a 25% increase due to inflation and the inefficiency of multipart construction:

1. After 1 year, add a cold storage warehouse capable of storing a million pounds. Costs and some details can be found in Kolbe et al. 2006. Refrigeration adjustments would be to add another compressor (60-75 HP) and three coil units in the cold room and up-size the condenser, tanks, and piping. Additional costs might be

Equipment
FOB points of manufacture $165,000

Piping and vessel insulation allowance $35,000

Freight $9,000

Installation $58,000

Under-floor heat grid allowance $35,000

Total $302,000

Not counted here is the cost of the cold storage building (materials, shipping, installation).

2. After a second year, add an ice plant capable of producing 30 tons of flake ice per day. The attached icehouse would store 100 tons, or a 3-day supply.

Refrigeration adjustments would be to add another 75 HP compressor and the icemaker, and up-size the condenser, tanks, and piping. Additional ballpark costs might be

Equipment
FOB points of manufacture	$200,000
Machine room components	$10,000
Freight	$17,000
Installation	$67,000
Mechanical ice delivery rake system, installed	$150,000
Total	**$444,000**

Not included is the cost of the insulated building housing the ice storage and delivery system.

ONBOARD BLAST FREEZER FOR ALBACORE TUNA

Background

A West Coast fisherman will convert a small, 56-foot inshore dragger into an albacore tuna troller. Targeting a market for high-quality frozen tuna, refrigeration and a freezing system will be added to this boat that previously carried iced fish. Low operating temperatures will require the surfaces of the 1,500 cubic foot hold to be insulated and lined.

The operation will be designed to handle up to 300 fish per day, with a size range from 12 to 24 pounds per fish. The fishing day can be dawn to dusk and the freezer must complete a freezing cycle by the start of a new day. (The freezer may have to operate all night when fishing is good.) Because the core temperature of the landed, struggling fish may be as high as 90°F, some way to prechill the catch on deck will be required for this high-quality product. Thus, the fish loaded into the freezer will be under 60°F. Freezing will bring the average fish temperature to a level approaching –20°F, the target temperature for the fish storage area.

Fishing opportunities beyond albacore may include longlined blackcod, halibut, or rockfish. The owner would like to have the system capable of freezing these species as well, when markets are available.

The Plan

The market dictates that blast-frozen, glazed albacore be off-loaded with a core temperature no higher than 0°F. If one assumed the average fish weight to be 17 pounds, then the refrigeration must be able to freeze 5,100 pounds per day. Experiments show that freezing times of prechilled albacore will be on the order of 8 hours (12-pound fish) to 12 hours (24-pound fish), and the space for hanging, shelving, glazing, and stacking in the hold must be suitably configured. Additionally, the boat will install an RSW system to chill two deck tanks having a total volume of around 90 cubic feet (Kolbe et al. 2004b).

The Major Pieces

Fish hold insulation
Spray and shape 6 inches of foam insulation over all surfaces of the 1,500 cubic foot hold (Area = 1,150 square feet).
Assume $1.30 per square foot per inch thickness.

($1.30/ft²-in) × (6 in) × (1,150 ft²) =	$9,000

Fiberglass liner
Install on all surfaces; assume $15 per square foot.

($15/ft²) × (1,150 ft²) =	$17,000

Four-ton RSW condenser unit

To include pumps, titanium chiller, controls	$14,000
Installation	$2,000

This is a module that includes both the refrigeration condensing unit, and a shell-and-tube barrel chiller. Its capacity will be 4 TR at 25°F evaporating temperature.

Blast freeze unit

To include refrigeration capable of 5,000 pounds of fish per day	$40,000
Installation	$15,000

Two evaporator coils are hung in the fish hold, plumbed to a refrigeration condensing unit located in the engine room. The two-stage R-507 system has a motor-driven compressor, and seawater-cooled condenser. It is capable of supplying 5 TR at –40°F evaporator temperature.

Total of major items	**$97,000**

The above assumes the owner will install racks and fish-handling equipment, onboard tanks, and plumbing. It does not include electric generator sets required to drive the blast freezer (15-20 kW) and the RSW unit (5 kW). Note that these are steady-state power requirements; starting loads will require a greater generating capacity.

Operation cost (besides maintenance) will include the fuel to run the generator sets (around 1 gallon per hour) and most significantly, extra labor needed to handle/process the fish for high-quality delivery.

Discussion

Given the long freezing times of whole tuna, fish caught late in the day may take all night to freeze. Although temptation might be to shut down at night due to noise, that may be impossible if the machinery is to keep pace with the catch rate.

Particularly after loading wet product into the freezer, frost forming on the evaporator fins will begin to shut off the airflow through the passages. It is up to the operator to manually initiate a de-frost cycle when needed. During this period, the fans switch off, electric heaters melt the frost into a pan, and the water drains out of the cold room.

Because a single blast freeze unit will likely control temperature throughout the hold, fish already frozen may lie in high velocity air streams. A good glaze must be applied to prevent dehydration in these conditions.

SHELF FREEZER FOR SPECIALTY PRODUCTS

Background

A coastal retail fish market currently sells unfrozen fish and shellfish from their facility housing a small, chilled storage room and display case. A new business plan calls for expansion to include a line of high-quality frozen packaged portions. Besides retail sales, markets might include restaurants, other food markets, and possibly mail order. A small walk-in cold storage room has been ordered; a new freezer must be selected and installed.

Initial products will be salmon fillets and steaks, and albacore tuna medallions and loins. These will be vacuum-packaged or packaged in trays, with portion sizes ranging from 4 to 10 ounces. Future products are expected to include a large range of species, sizes, and preparations; examples are breaded and seasoned rockfish fillets.

A production target is 5,000 pounds of round product per day, delivered to the plant and pro-cessed within a 10-hour day.

The Plan

A need for rapid freezing, anticipated product size-variation, limited plant space, and desired low cost will be factors. A contractor has recommended that a stationary shelf freezer be considered. The type of products and yields will affect the freezer load. If steaks are cut from albacore tuna and salmon, re-coveries of around 60% can be expected; skinless/boneless fillets of rockfish, salmon, and albacore will be in the range of 30-35% (Crapo et al. 2004).

The freezer will be sized for the higher yield (60%) from the incoming round product; thus 3,000 pounds must be frozen per 9-hour period. Representative samples of packaged 6-ounce salmon steaks were sent to a shelf-freezer manufac-turer for trials. Freezing times ranged from 30 to 50 minutes, depending on the several packaging options used. If freezing cycles, to include loading and unloading times, are to be 1.5 hours, the freez-er shelf area must support 500 pounds of products, or 1,333 packages of 6-ounce portions.

On occasion, process delays or markets may require dressed salmon, halibut, or rockfish to be frozen whole. These could be loaded at the end of the shift for overnight freezing. Greater spacing between shelves would have to accommodate these alternative products.

The Major Pieces

Two 10-station freezer racks with 11 plates, measuring 32 x 90 inches; unit price is $7,000 each   **$14,000**

Aluminum plates have fins on lower surface to enhance heat transfer to the surrounding air. Plate spacing for 7 of the 10 stations is 3 inches; the remainder are 5 inches (for larger fish and cartons).

Discussion

The contractor's listing includes the freezer shelves, frames, and piping. New or existing refrigeration equipment, and a cold room or insulated cabinet, would have to be supplied by the customer in this case. Cabinets are typically constructed with 3 inches of foam insulation and two opening doors. When housed within an insulated cabinet, a common option is a centrifugal fan that moves inside air to speed freezing.

CRYOGENIC FREEZER FOR OYSTERS

Background

A shellfish grower with a small enterprise sees an opportunity to add product value by setting up a small processing plant. The major farmed product is oysters—some bottom-cultured and the rest grown on racks. The rack-cultured oysters grow as individuals (vs. clusters), targeted to the half-shell trade.

A market has appeared for individually quick-frozen (IQF) half-shell oysters packaged in trays, then packed in cartons, 12 dozen per carton. The buyer requires fast-frozen, high-quality products, and has mentioned cryogenics as an advantage. Besides freezing fast, there are indications that pathogen levels are reduced as the result of cryo-genic freezing with CO_2.

While retaining some of the production for the fresh oyster trade, the grower projects that about 20 bushels per day could be directed to the frozen half-shell market. His production plan will be to operate one 8-hour shift per day, 5 days a week. Oyster production could go 8 months per year. As this changes or fluctuates, the grower sees other op-portunities for such products as breaded IQF oyster meats, half-shell scallops, and prawns.

The Plan

Production rate of frozen half-shells will be on the order of 900 oysters per hour over an 8-hour shift. The oysters with upper shell removed will be frozen, glazed, packed into trays (12 per tray) then into cartons (12 trays per carton), which are then transported to frozen storage. The target is 50 cases per day.

With the top shell removed, the shucked oysters will weigh on the order of 0.4 pounds. Calculations anticipate the minimum refrigeration capacity to be 3.6 refrigeration tons. While an air-blast tunnel freezer could easily achieve this rate, the plan will be to install a carbon dioxide cryogenic freezing system. Experiments determine freezing time in the –109°F gas to be about 10 minutes.

Either a cabinet freezer or tunnel could do the job. The vendor recommends a small tunnel freezer because of its higher efficiency in the use of carbon dioxide.

The storage tank will be leased from the gas company; the processor will install supply piping and exhaust ducts, power supply, and foundations.

The Major Pieces

Tunnel freezer is a 10-foot section with a 26-inch-wide continuous belt feed. $52,000

Piping. A length of 85 feet of urethane-insulated piping will connect the freezer with a storage tank immediately outside the building. A rough estimate for this piping is $65 per foot. $6,000

Total capital costs $58,000

Some monthly charges

Storage tank. Tanks are typically leased to the user by the gas supplier. For this installation, typical cost might be $750 per month

Gas use. The vendor estimates a use rate of 1.5 pounds of carbon dioxide per pound of product frozen. This will include heat leakage, initial cool-down of the tunnel, belt cooling. Cost of liquid carbon dioxide is estimated at $100 per ton. $4,800 per month

Discussion

A few charges will immediately add to the above costs: foundations for both the freezer and the stor-age tank, built to proper specs; an exhaust system for the cold gas leaving the freezer; and electric power (and its monthly charge) to drive the con-veyor and fans.

One alternative to the carbon dioxide tunnel configuration would be a single-door cabinet freezer. Oysters placed on 19-shelf racks would be wheeled into the cabinet. With 10-minute freeze times, three cycles per hour would easily handle the required production. The freezer, including three sets of racks, would cost around $42,000. This is $10,000 less than the tunnel. Its disadvan-tage is in its higher rate of gas use—2.5 pounds per pound of product—due to its lower efficiency of use. The lower gas use for the tunnel freezer would pay back the higher capital cost of the tunnel within about 3 months.

It is not clear that carbon dioxide freezing is more effective than other freezing methods in killing off pathogens. Kilgen and Hernard (1995), investigating reduction of *Vibrio vulnificus* in Gulf Coast oysters, found blast freezing and carbon dioxide freezing to be equally effective. So the advantage of cryogenics may be, as on the list presented earlier, potentially better freezing quality with faster freezing rates, flexibility of freezing other products at different rates, and lower capital cost of equipment. Depending on the geographic area and availability of cryogens, it is likely that liquid nitrogen would be another good alternative.

References

Ablett, R.F., and S.P. Gould. 1986. Comparison of the sensory quality and oxidative rancidity status of frozen-cooked mussels (*Mytilus edulis* L.). J. Food Sci. 51(3):809-811.

Ablett, R.F., S.P. Gould, and D.A. Sherwood. 1986. Frozen storage performance of cooked cultivated mussels (*Mytilus edulis* L.): Influence of ascorbic acid and chelating agents. J. Food Sci. 51(5):1118-1121.

Archer, D.L. 2004. Freezing: An underutilized food safety technology. Int. J. Food Microbiology 90(2):127-138.

ASHRAE. 1994. Refrigeration handbook. ASHRAE (American Society of Heating, Refrigerating, and Air Conditioning Engineers), Atlanta.

ASHRAE. 1997. Fundamentals handbook. ASHRAE (American Society of Heating, Refrigerating, and Air Conditioning Engineers), Atlanta.

ASTM. 1981. Manual on the use of thermocouples in temperature measurement. Special Technical Publication 470 B. ASTM (American Society for Testing and Materials), Philadelphia.

Auh, J.H., H.G. Lee, J.W. Kim, I.C. Kim, H.S. Yoon, and K.H. Park. 1999. Highly concentrated branched oligosaccharides as cryoprotectant for surimi. J. Food Sci. 64(3):418-422.

Aurell, T., B. Dagbjartsson, and E. Salomonsdottir. 1976. A comparative study of freezing qualities of seafoods obtained by using different freezing methods. J. Food Sci. 41:1165-1167.

Balthrop, P. Undated. Ancient delicacy, modern technology. Florida Department of Agriculture and Consumer Services Information, Bureau of Seafood and Aquaculture Release. 2 pp.

Banks, A. J.A. Dassow, E.A. Feiger, A.F. Novak, J.A. Peters, J.W. Slavin, and J.J. Waterman. 1977. Freezing of shellfish. In: N.W. Derosier and D.K. Tressler (eds.), Fundamentals of food freezing. AVI Publishing Co., Westport, Connecticut, pp. 318-356.

Bell, L.N., and D.E. Touma. 1996. Glass transition temperatures determined using a temperature-cycling differential scanning calorimeter. J. Food Sci. 61(4):807-810 and 828.

Bilinski, E., R.E.E. Jonas, Y.C. Lau, and G. Gibbard. 1977. Treatments before storage affecting thaw drip formation in Pacific salmon. J. Fish Res. Board Can. 34:1431-1435.

Brake, N.C., and O.R. Fennema. 1999. Glass transition values of muscle tissue. J. Food Sci. 64(1):10-15.

Chevalier, D., M. Sentissi, M. Havet, and A. Le Bail. 2000. Comparison of air-blast and pressure shift freezing on Norway lobster quality. J. Food Sci. 65(2):329-333.

Cieloha, R. 2005. Vice president, Permacold Engineering, personal communication.

Claggett, T.J., W.A. Clayton, B.G. Liptak, and R.W. Worrall. 1982. Temperature measurements. Chapter 4. In: B.G. Liptak and K. Venczel (eds.), Instrument engineeers' handbook. Chilton Book Co., Radnor, Pennsylvania.

Clark, J.P. 1997. Cost and profitability estimation. In: K.J. Valentas, E. Rotstein, and R.P. Singh (eds.), Handbook of food engineering practice. CRC Press, Boca Raton.

Cleland, A.C., and R.L. Earle. 1984. Freezing time prediction for different final product temperatures. J. Food Sci. 49:1230-1232.

Cleland, D.J., and K.J. Valentas. 1997. Prediction of freezing time and design of food freezers. Chapter 3. In: K.J. Valentas, E. Rotstein, and R.P. Singh (eds.), Handbook of food engineering practice. CRC Press, Boca Raton.

Crapo, C., B. Paust, and J. Babbitt. 2004. Recoveries and yields from Pacific fish and shellfish. Alaska Sea Grant, University of Alaska Fairbanks.

Craven, C., E. Kolbe, M. Morrissey, and G. Silvia. 1997. Onboard factors affecting chilling and freezing rates, and quality of albacore tuna. A report to the American Fishermen's Research Foundation. Coastal Oregon Marine Experiment Station, Oregon State University, Newport.

DeBeer, J. 1998. Accurately measuring seafood temperatures. Available at seafood.ucdavis.edu/Pubs/tempdoc.htm. (Accessed Jan. 2007.)

Denys, S., A.M. Van Loey, M.E. Hendricks, and P.P. Tobback. 1997. Modeling heat transfer during high-pressure freezing and thawing. Biotechnol. Prog. 13(4):416-422.

Dewey, M. 2001. Compare real vs. initial cost of cryogenic piping. Process Cooling and Equipment, March-April, pp. 25-29.

Dorgan, C.B., S.P. Leight, and C.E. Dorgan. 1995. Application guide for absorption cooling refrigeration using recovered heat. ASHRAE (American Society of Heating, Refrigerating, and Air Conditioning Engineers), Atlanta.

Doyle, J.P., and C. Jensen. 1988. Handbook on whitefish handling aboard fishing vessels. Alaska Sea Grant, University of Alaska Fairbanks.

Elenbaas, C.J. 1989. Food program manager. Liquid Air Corp., personal communication.

Energy Concepts Co. 1994. Advanced absorption refrigeration from waste heat. Final project report for Project 91-2-100. Alaska Science and Technology Foundation, Anchorage.

Erickson, D. 1996. Refrigeration with waste heat. In: Proceedings of the 18th Annual Meeting of the International Institute of Ammonia Refrigeration, Atlanta, pp. 21-31.

Fennema, O. 1973. Nature of the freezing process. Chapter 4. In: O. Fennema, W.D. Powrie, and E.H. Marth (eds.), Low-temperature preservation of foods and living matter. Marcel Dekker, New York.

Fennema, O. 1975. Freezing preservation. Chapter 6. In: M. Karel, O.R. Fennema, and D.B. Lund (eds.), Physical principles of food preservation. Marcel Dekker, New York.

Fikiin, K. 2003. Novelties of food freezing research in Europe and beyond. A presentation from the European Commission Flair-Flow 4, Institut National de la Recherche Agronomique, Paris. 56 pp. Available at www.nutrition.org.uk/upload/SME%209%20foodfreeze.pdf. (Accessed Jan. 2007.)

Gameiro, W. 2002. Energy costs are changing refrigeration design. Technical Paper No. 6, Proceedings of the 2002 IIAR Ammonia Refrigeration Conference, Kansas City, Missouri, pp. 207-233.

Garthwaite, G.A. 1997. Chilling and freezing of fish. Chapter 4. In: G.M. Hall (ed.), Fish processing technology. 2nd edn. Blackie Academic and Professional, London.

Gibbard, G. 1978. Freezing seafood at sea: The British Columbia experience. In: J. Peters (ed.), Seafood handling, preservation, marketing. Proceedings of a technical conference. University of Washington Sea Grant, Seattle, pp. 51-65.

Gibbard, G.A., and S.W. Roach. 1976. Standards for an RSW system. Environment Canada, Fisheries and Marine Service Technical Report No. 676.

Gibbard, G.A., F. Lee, S. Gibbard, and E. Bilinski. 1982. Transport of salmon over long distances by partial freezing in RSW vessels. In: Proceedings of the International Institute of Refrigeration Conference on Advances in the Refrigerated Treatment of Fish, Boston, 1981.

Goff, H.D. 1997. Measurement and interpretation of the glass transition in frozen foods. In: M.C. Erickson and Y.-C. Hung (eds.), Quality in frozen food. Chapman and Hall, New York, pp. 29-50.

Graham, J. 1974. Installing an air-blast freezer? Torry Advisory Note No. 35. Torry Research Station, Aberdeen.

Graham, J. 1977. Temperature and temperature measurement. Torry Advisory Note No. 20. Torry Research Station, Aberdeen.

Graham, J. 1984. Planning and engineering data. 3. Fish freezing. FAO Fisheries Circular No. 771. Available at www.fao.org/DOCREP/003/R1076E/R1076E00.htm. (Accessed Jan. 2007.)

Haard, N.F. 1992. Biochemical reactions in fish muscle during frozen storage. In: E.G. Bligh (ed.), Seafood science and technology. Fishing News Books, Cambridge, pp. 176-209.

Hall, G.M., and N.H. Ahmad. 1997. Surimi and fish-mince products. In: G.M. Hall (ed.), Fish processing technology. 2nd edn. Chapman and Hall, London, pp. 74-92.

Heldman, D.R. 1992. Food freezing. Chapter 6. In: D.R. Heldman and D.B. Lund (eds.), Handbook of food engineering. Marcel Dekker, New York.

Heldman, D.R., and R.P. Singh. 1981. Food process engineering. 2nd edn. AVI Publishing Co., Westport, Connecticut.

Hilderbrand, K.S. 1979. Preparation of salt brines for the fishing industry. Oregon State University Sea Grant Extension, Newport.

Hilderbrand, K.S. 2001. Sea Grant Extension seafood technologist, Oregon State University, personal communication.

Howgate, P. 2003. UK/Torry seafood technologist, retired, email communication.

Hung, Y.-C., and N.-K. Kim. 1996. Fundamental aspects of freeze-cracking. Food Technol. 50(12):59-61.

Hurling, R., and H. McArthur. 1996. Thawing, refreezing and frozen storage effects on muscle functionality and sensory attributes of frozen cod (*Gadus morhua*). J. Food Sci. 61(6):1289-1296.

Jittinandana, S., P.B. Kenney, and S.D. Slider. 2005. Cryoprotectants affect physical properties of restructures trout during frozen storage. J. Food Sci. 70(1):C35-C42.

Johnston, W.A., F.J. Nicholson, A. Roger, and G.D. Stroud. 1994. Freezing and refrigerated storage in fisheries. FAO Fisheries Technical Paper 340. 143 pp. Available at www.fao.org/DOCREP/003/V3630E/V3630E00.htm. (Accessed Jan. 2007.)

Jul, M. 1984. The quality of frozen foods. Academic Press, Orlando, Florida. 292 pp.

Kallenberg, W.A. 2003. Alternative refrigeration strategies for rural Alaska. In: Proceedings of the conference A Public Seafood Processing and Cold Storage Facility: Is It Right for Your Community? November 2003, Anchorage. Available at www.uaf.edu/MAP/workshops/cold-storage/index.html. (Accessed Jan. 2007.)

Kilgen, M.B., and M.T. Hernard. 1995. Evaluation of commercial irradiation and other processing methods for *Vibrio vulnificus* control in Louisiana oysters. Presented to the Seafood Science and Technology Society of the Americas, November 5-9, 1995, Humacao, Puerto Rico.

Kim, N.-K., and Y.-C. Hung. 1994. Freeze-cracking in foods as affected by physical properties. J. Food Sci. 59(3):669-674.

Kolbe, E. 1990. Estimating energy consumption in surimi processing. J. Appl. Engineering Ag. 6(3):322-328.

Kolbe, E. 1991. An interactive fish freezing model compared with commercial experience. Proceedings of the 18th International Congress of Refrigeration, August 1991, Montreal, Canada. Available at seagrant.oregonstate.edu/extension/fisheng/freeze.html. (Accessed Jan. 2007.)

Kolbe, E., and D. Cooper. 1989. Monitoring and controlling performance of commercial freezers and cold stores. Final report submitted to Alaska Fisheries Development Foundation, Project NA-89-ABH-00008, "Groundfish Quality Enhancement."

Kolbe, E., D. Kramer, and J. Junker. 2006. Planning seafood cold storage. 3rd edn. Alaska Sea Grant, University of Alaska Fairbanks. 78 pp.

Kolbe, E., Q. Ling, and G. Wheeler. 2004a. Conserving energy in blast freezers using variable frequency drives. In: Proceedings of the 26th National Industrial Energy Technology Conference, Houston, pp. 47-55.

Kolbe, E., C. Craven, G. Sylvia, and M. Morrissey. 2004b. Chilling and freezing guidelines to maintain onboard quality and safety of albacore tuna. Oregon State University Agriculture Experiment Station Special Report 1006.

Krack Corp. 1992. Refrigeration load estimating manual. RLE-593. Krack Corp., Addison, Illinois. 52 pp.

Kramer, D.E., and M.D. Peters. 1979. Utilization of Pacific rockfish. 3. A quality comparison of *Sebastes alutus* and *Sebastes flavidus* during chill and frozen storage. Fish. Res. Board Can. Tech. Rep. 879. 54 pp.

Kreuzer, R. (ed.). 1969. Freezing and irradiation of fish. Proceedings FAO Conference, September 1967, Madrid. Fishing News (Books), Ltd., London.

Krzynowek, J., and K. Wiggin 1979. Seasonal variation and frozen storage stability of blue mussels (*Mytilus edulis*). J. Food Sci. 44(6):1644-1645 and 1648.

Le Meste, M., D. Champion, G. Roudaut, G. Blond, and D. Simatos. 2002. Glass transition and food technology: A critical appraisal. J. Food. Sci. 67(7):2444-2458.

Levine, H., and L. Slade. 1989. Response to the letter by Simatos, Blond, and Le Meste on the relation between glass transition and stability of a frozen product. Cryo-Letters 10:347-370.

Li, B., and D.W. Sun. 2002a. Effect of power ultrasound on freezing rate during immersion freezing of potatoes. J. Food Eng. 55:277-282.

Li, B., and D.W. Sun. 2002b. Novel methods for rapid freezing and thawing of foods: A review. J. Food Eng. 54:175-182

Licciardello, J.J. 1990. Freezing. In: R.E. Martin and G.J. Flick (eds.), The seafood industry. Van Nostrand, New York, pp. 205-218.

Lindsey, S. 2005. Principal engineer, Western Region, BOC Gases, personal communication.

Löndahl, G., and S. Goransson. 1991. Influences of different glazing techniques on the quality of shellfish and shellfish products. In: Proceedings of the XVIII International Congress of Refrigeration, Montreal.

Love, R.M. 1966. The freezing of animal tissue. In: H.T. Meryman (ed.), Cryobiology. Academic Press, New York, pp. 317-405.

Love, R.M. 1988a. Texture. In: The food fishes. Van Nostrand Reinhold Co., New York, pp. 121-140.

Love, R.M. 1988b. Gaping. In: The food fishes. Van Nostrand Reinhold Co., New York, pp. 161-180.

Magnússon, H., and E. Martinsdóttir. 1995. Storage quality of fresh and frozen-thawed fish in ice. J. Food Sci. 60(2):273-278.

Mannapperuma, J.D., and R.P. Singh. 1988. Prediction of freezing and thawing times of foods using a numerical method based on enthalpy formulation. J. Food. Sci. 53(2):626-630.

Mannapperuma, J.D., and R.P. Singh. 1989. A computer-aided method for the prediction of properties and freezing/thawing of foods. J. Food Eng. 9(1989):275-304.

Martino, M.N., L. Otero, P.D. Sanz, and N.E. Zaritzky. 1998. Size and location of ice crystals in pork frozen by high-pressure-assisted freezing as compared to classical methods. Meat Science 50(3):303-313.

Martinsdóttir, E., and H. Magnússon. 2001. Keeping quality of sea-frozen thawed cod fillets on ice. J. Food Sci. 66(9):1402-1408.

Merritt, J. 1982. Professor, Technical University of Nova Scotia, personal communication.

Merritt, J.H. 1978. Refrigeration on fishing vessels. Fishing News Books Ltd., Farnham, England.

References

Ming, Z. 1982. Applications of partial freezing technique on fishing vessels operating in the South China Sea. In: Proceedings of the International Institute of Refrigeration Conference on Advances in the Refrigerated Treatment of Fish, 1981, Boston.

Moosavi-Nasab, M., I. Alli, A.A. Ismail, and M.O. Ngadi. 2005. Protein structure changes during preparation and storage of surimi. J. Food Sci. 70(7):448-453.

Morrison, C.R. 1993. Fish and shellfish. In: C.P. Mallett (ed.), Frozen food technology. Chapman and Hall, London, pp. 196-236.

Morrison, G.S., and F.P. Veitch. 1957. An investigation of the chemistry of texture changes of frozen blue crab meat. Commer. Fish. Rev. 19(10):1-5.

Morrissey, M. 2005. Director, Oregon State University Seafood Laboratory, personal communication.

Mousavi, R., T. Miri, P.W. Cox, and P.J. Fryer. 2005. A novel technique for ice crystal visualization in frozen solids using x-ray micro-computed tomography. J. Food Sci. 70(7):437-442.

Nesvadba, P. 1993. Glass transitions in aqueous solutions and foodstuffs. In: J.M.V. Blanshard and P.J. Lillford (eds.), The glassy state in foods. Nottingham University Press, Loughborough, UK, pp. 523-526.

Nesvadba, P. 2003. Frozen food properties. In: D. Heldman (ed.), Encyclopedia of agricultural, food, and biological engineering. Marcel Dekker, New York.

Nicholson, F.J. 1973. The freezing time of fish. Torry Advisory Note 62. 7 pp.

Nilsson, K., and B. Ekstrand. 1994. Refreezing rate after glazing affects cod and rainbow trout muscle tissue. J. Food Sci. 59(4):797-798 and 838.

Nordin, D.M.A., and D.E. Kramer. 1979. Studies on the handling and processing of sea urchin roe. II. Frozen product. Fish. Res. Board Can. Tech. Rep. 877. 19 pp.

Ocean Leader. 1981. Cryogenic freezing: −300°F is too cold too fast for larger fish. Ocean Leader 1(1):41-42.

Ogawa, Y. 1988. Freezing system with thermal equalizing. Japan Refrig. 63:40-46.

Omega Engineering. 1998. Bluecat buyers guide book 1. Stamford, Connecticut.

Park, J.W. 1994. Cryoprotection of muscle proteins by carbohydrates and polyalcohols: A review. J. Aquat. Food Prod. Technol. 3(3):23-41.

Persson, P.O., and G. Löndahl. 1993. Freezing technology. Chapter 2. In: C.P. Mallett (ed.), Frozen food technology. Blackie Academic and Professional, London.

Pham, Q.T., and R.F. Mawson. 1997. Moisture migration and ice recrystallization in frozen foods. Chapter 5. In: M.C. Erickson and Y.C. Hung (eds.), Quality in frozen food. Chapman and Hall, New York.

Piho, J. 2000. Notes by Refrigeration Service and Design, Inc., West Conshohocken, Pennsylvania. Distributed at the Industrial Refrigeration Workshop, held in Charlotte, North Carolina. Sponsored by Kansas State University Continuing Education, Manhattan, Kansas.

Préstamo, G., L. Palomares, and P. Sanz. 2005. Frozen foods treated by pressure shift freezing: Proteins and enzymes. J. Food Sci. 70(1):S22-S27.

Price, R.J., D.E. Kramer, and P.D. Tom. 1996. Processing mussels: The HACCP way. University of California Sea Grant Extension Program, UCSGEP 96-2. 43 pp.

Rahman, S. 1995. Food properties handbook. CRC Press, Boca Raton.

Rasmussen, D. 1969. A note about "phase diagrams" of frozen tissue. Biodynamica 10:333-339.

Regenstein, J.M., M.A. Schlosser, A. Samson, and M. Fey. 1982. Chemical changes in trimethylamine oxide during fresh and frozen storage of fish. In: R.E. Martin, G.J. Flick, C.E. Hebard, and D.R. Ward (eds.), Chemistry and biochemistry of marine food products. AVI Publishing Co., Westport, Connecticut, pp. 137-148.

Reid, D.S. 1990. Optimizing the quality of frozen foods. Food Technol. 44(7):78-82.

Reid, D.S. 1993. Basic physical phenomena in the freezing and thawing of plant and animal tissues. In: C.P. Mallett (ed.), Frozen food technology. Chapman and Hall, London.

Reid, D.S. Undated. The significance of the glassy state in the storage of frozen foods. Unpublished. 2 pp.

Reid, D.S., N.F. Doong, M. Snider, and A. Foin. 1986. Changes in the quality and microstructure of frozen rockfish. In: D.E. Kramer and J. Liston (eds.), Seafood quality determination. Elsevier Science Publishers, Amsterdam, pp. 1-15.

Rørå, A.M.B., and O. Einen. 2003. Effects of freezing on quality of cold-smoked salmon based on measurements of physiochemical characteristics. J. Food Sci. 68:2123-2128.

Ruan, R., Z. Long, P. Chen, V. Huang, S. Almaer, and I. Taub. 1999. Pulse NMR study of glass transition in maltodextrin. J. Food Sci. 64(1):6-9.

Rudolph, A.S., and J.H. Crowe. 1985. Membrane stabilization during freezing: The role of two natural cryoprotectants, trehalose, and proline. Cryobiology 22:367-377.

Salvadori, V.O., and R.H. Mascheroni. 2002. Analysis of impingement freezers performance. J. Food Eng. 54(2002):133-140.

Sathivel, S. 2005. Chitosan and protein coatings affect yield, moisture loss, and lipid oxidation of pink salmon (*Oncorhynchus gorbuscha*) fillets during frozen storage. J. Food Sci. 70(8):455-459.

Schlimme, D.V. 2001. Rapid pressure-shift freezing technique yields non-frozen-like food texture. Quick Frozen Foods International, July, p. 168.

Sikorski, Z.E., and A. Kolakowska. 1990. Freezing of marine food. In: Z.E. Sikorski, Resources, nutritional composition, and preservation. CRC Press, Boca Raton, pp. 111-124.

Simatos, D., and G. Blond. 1993. Some aspects of the glass transition in frozen foods systems. In: J.M.V. Blanshard and P.J. Lillford (eds.), The glassy state in foods. Nottingham University Press, Loughborough, UK, pp. 395-415.

Simatos, D., M. Faure, E. Bonjour, and M. Couach. 1975. Differential thermal analysis and differential scanning calorimetry in the study of water in foods. In: R.B. Duckworth (ed.), Water relations of foods. Academic Press, London, pp. 193-209.

Singh, R.P., and J.D. Mannapperuma. 1990. Developments in food freezing. Chapter 11. In: H.G. Schwartzberg and M.A. Rao (eds.), Biotechnology and food process engineering. Marcel Dekker, New York.

Slutskaya, T.N. 1973. Effect of freezing on the nutritive value of Echinodermata. Issled. Technol. Rybn. Prod. 4:16-22.

Sternin, V. 1991. Possible adverse effect of cryogenic freezing. Paper presented at the 42nd Annual Meeting of the Pacific Fisheries Technologists, Feb. 17-20, Victoria, B.C., Canada.

Sternin, V., and I. Dore. 1993. Caviar: The resource book. Cultura, Moscow, Russia. 256 pp.

Stoecker, W.F. 1998. Industrial refrigeration handbook. McGraw Hill. New York.

Storey, R.M. 1980. Modes of dehydration of frozen fish flesh. In: J.J. Connell and Torry Research Station staff (eds.), Advances in fish science and technology. Fishing News Books Ltd., Farnham, Surrey, UK, pp. 498-502.

Sun, D.W., and B. Li. 2003. Microstructural change of potato tissues frozen by ultrasound-assisted immersion freezing. J. Food Eng. 57:337-345.

Sutton, R.L., I.D. Evans, and J.F. Crilly. 1994. Modeling ice crystal coarsening in concentrated disperse food systems. J. Food Sci. 59(6):1227-1233.

Takeko, H. 1974. Processing and refrigeration facilities for a fish factory ship. Proceedings of the International Institute of Refrigeration 1:203-213.

Tanaka, K., and S.T. Jiang. 1977. Freezing preservation of boiled and fresh sea cucumbers (*Stichopus japonicus* Selenka). Refrigeration JP. 52(599):789-792.

Tomás, M.C., and M.C. Añón. 1990. Study on the influence of freezing rate on lipid oxidation in fish (salmon) and chicken breast muscles. Int. J. Food Sci. Technol. 25(6):718-721.

Tomlinson, N., S.E. Geiger, D.E. Kramer, and S.W. Roach. 1973. The keeping quality of frozen halibut in relation to prefreezing treatment. Fish. Res. Board Can. Tech. Rep. 363. 21 pp.

USDA. 2001. Parasites. In: Fish and fisheries products hazards and controls guidance. 3rd edn. U.S. Food and Drug Administration Office of Seafood, Washington, D.C., pp. 65-71.

Wang, D.Q., and E. Kolbe. 1989. Prediction of heat leakage through fish hold wall sections. J. Ship Research 33(3):229-235.

Wang, D.Q., and E. Kolbe. 1990. Thermal conductivity of surimi: Measurement and modeling. J. Food Sci. 55(5):1217-1221, 1254.

Wang, D.Q., and E. Kolbe. 1991. Thermal properties of surimi analyzed using DSC. J. Food Sci. 56(2):302-308.

Wang, D.Q., and E. Kolbe. 1994. Analysis of food block freezing using a PC-based finite element package. J. Food Eng. 21(1994):521-530.

Warwick, J. 1984. A code of practice for mussel processing. New Zealand Fishing Industry Board. 35 pp.

Watanabe, H., C.Q. Tang, T. Suzuki, and T. Mihori. 1996. Fracture stress of fish meat and the glass transition. J. Food Eng. 29:317-327.

Waterman, J.J., and D.H. Taylor. 1967. Superchilling. Torry Advisory Note No. 32. Torry Research Station, Aberdeen.

Weiner, A. 2003. Your passage into cryogenic pipe types. Process Cooling and Equipment. Jan.-Feb., pp. 28-29.

Whited, R. 2005. Sales engineer, Air Products, Inc., personal communication.

Wilcox, M.H. 1999. State-of-the-art energy efficiency in refrigerated warehouses. Proceedings of the IIAR (International Institute of Ammonia Refrigeration), Dallas, Texas. Technical Paper 15, pp. 327-340.

World Food Logistics Organization. 2002. Industrial-scale food freezing: Simulation and process. Version 3.0. World Food Logistics Organization, 1500 King St., Suite 201, Alexandria, VA 22314.

Wray, T. 2005. Freezing thin products fast. Seafood Processor 2(12):14-15.

Zhao, Y., E. Kolbe, and C. Craven. 1998. Simulation of onboard chilling and freezing of albacore tuna. J. Food Sci. 65(5):751-755.

Zhu, S., A. Le Bail, H.S. Ramaswamy, and N. Chapleau. 2004. Characterization of ice crystals in pork muscle formed by pressure-shift freezing as compared with classical freezing methods. J. Food Sci. 69(4):FEP190-FEP197.

❄❄❄❄❄❄❄❄❄

Appendix

DEFINITIONS, UNITS, CONVERSIONS

In engineering, there are two systems of units, the **English System**, also called the "I-P" (inch-pound) system; and the **SI System** (System Internationale), a version of the metric system.

Energy

Also heat or work.

English: Btu (British thermal unit)
It takes 1 Btu of heat to change 1 pound of water 1°F.

SI: J (joule) or kJ (kilojoule)
Also, kWh (kilowatt-hour: 1 kW of power operating for 1 hour.)

1 Btu = 1.055 kJ

Power

Rate of transferring energy; rate of doing work.

English: Btu per hr (Btuh)
Horsepower (HP)
Refrigeration ton (TR). Used to describe the rate of heat absorbed by a refrigeration system. It is equivalent to a ton of ice melting in 24 hours.

SI: W (watts) or kW (kilowatts)

Note on electrical terminology: "Ohm's Law" is
Volts = Amps × Ohms
Electrical power = Volts × Amps = Amps² × Ohms = Watts

1 W = 1 Joule per second

1 HP = 0.7457 kW

1 Btu per hour = 0.2931 W

1 TR = 12,000 Btu per hour = 3.52 kW = 4.72 HP

Mass

English: lb (pound), sometimes noted "lbm" for "pound mass," vs. "lb" for "pound force"
t (ton), also called "short ton"

SI: kg (kilogram)
t (ton) (usually called "metric ton")

1 short ton = 2,000 lbm

1 metric ton = 1,000 kg

1 kg = 2.2 lbm

1 lbm = 0.45359 kg

Temperature

English: °F (degrees Fahrenheit)

SI: °C (degrees centigrade, Celsius)

A change of 1°C is equivalent to a change of 1.8°F

0°C = 32°F

100°C = 212°F

°C = (°F − 32)/1.8

°F = (°C) × (1.8) + 32

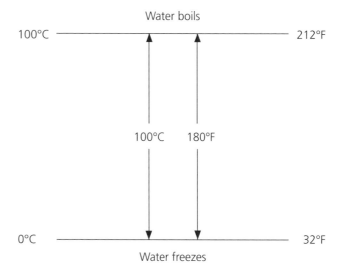

Pressure

English: psi (pounds per square inch)
psia (pounds per square inch absolute)
psig (pounds per square inch gage) in. Hg (inches of mercury)
A vacuum is measured in "inches of mercury vacuum" below atmospheric pressure.

SI: Pa (pascals)
kPa (kilopascals)

1 atm (1 atmosphere) = 101.3 kPa
1 atm = 14.7 psia
= 0 psig
= 29.92 in. Hg absolute pressure.
A pressure change of 1 psi = 2.035 in Hg

Pressure Ratio: The ratio of two absolute pressures. An example is pressure ratio of a refrigeration compressor for which
Pressure Ratio = (Compressor Discharge Abs. Pressure)/(Compressor Suction Abs. Pressure)
= (High Pressure Side/Low Side Pressure)

Length

English: ft (foot)

SI: m (meter)

1 ft = 0.3048 m

Volume

English: ft³ (cubic feet)
gal (gallons)

SI: m³ (cubic meters)
L (liter)

1 ft³ = 7.48 gal
1 m³ = 1,000 L
1 ft³ = 0.2832 m³
1 gal = 3.785 L

Thermal Conductivity

A quantity measuring how easily heat flows by conduction within a solid object.

English: k (Btu/hr-ft-F)
Sometimes, Btu-in/hr-ft²-F

SI: k (W/m-C)

1 Btu/hr-ft-F = 1.731 W/m-C

Heat Transfer Coefficient

A quantity measuring how easily heat flows by convection between a solid surface and a fluid.

English: U (Btu/hr-ft²-F)

SI: U (W/m²-C)

1 Btu/hr-ft²-F = 5.678 W/m²-C

Specific Heat

The amount of heat required to change a unit mass of a material by one degree.

English: Btu/lbm-F

SI: J/g-C or kJ/kg-C

For water, specific heat = 1

1 Btu/lbm-F = 4.1865 J/g-C = 4.1865 kJ/kg-C

Latent Heat of Fusion

The heat that is absorbed by a solid as it turns to liquid.

Ice absorbs 144 Btu/lbm as it melts.

Latent Heat of Vaporization

The heat that is absorbed by a liquid as it turns to vapor.

Water absorbs 970 Btu/lbm as it vaporizes or boils.

Latent Heat of Sublimation

The heat that is absorbed by a solid as it turns to vapor. Ice absorbs 1,114 Btu/lbm as it vaporizes.

TEMPERATURE MEASUREMENT ERRORS

As you use any of the temperature measurement devices described in Chapter 1, there are several things that will cause errors—and most relate to how the device is used.

Instrument Error

No sensor will give an **exact** correct reading, although thermistors and thermocouples are usually close—within a degree or two. Dial thermometers are notoriously inaccurate. Whichever sensor you're using, it is important to calibrate it often. For low temperature readings, calibrate in a mixture of ice and freshwater, preferably mixed together in a vacuum-walled thermos container (see also DeBeer 1998). Do this carefully, because the water can stratify into layers of different temperatures. The heaviest water is at 39.2°F (4°C) and it will sink to the bottom (Figure A-1). To calibrate, first adjust the reading on the meter if possible, or simply write down the difference between your reading and 32°F, and use that difference to correct all the readings that you take.

Misplacement

It is often tricky to place the sensor in the **exact** spot of interest. Figure 1-3 shows what happens if a probe is not located at the geometric center and the "last-point-to-freeze." Figure 1-17 indicates similar problems that result if the product itself is not freezing evenly, in which case your probe, though geometrically centered, is still not placed in the "last-point-to-freeze." Finally, especially in small products some error will result if you don't know where, inside the probe sheath or tube, the actual temperature sensing element is. We usually assume it to be right at the tip, but check that by gripping it between your fingers and noting the temperature jump on the meter.

Conduction Error

The rate of heat transfer (Q) by conduction through a solid can be described by the equation

$$\frac{Q}{A} = \frac{k}{l}\, \Delta T \quad \text{(Btu/hr-ft}^2\text{)}$$

where

A is the cross-sectional area of the heat flow path

l is the length of the heat flow path between T_{HOT} and T_{COLD}

ΔT is $T_{HOT} - T_{COLD}$

k is "thermal conductivity," a property of the material through which the heat is flowing.

For the walls of a stainless steel probe, k is a big number (such as 8) compared to that of frozen fish (0.8) or unfrozen fish (0.28). So when a metal probe is inserted into a fish, as in Figure A-2, penetration (l in our equation) must be deep. If it is not, heat will flow rapidly from the warm air (T_{HOT}) to the

Figure A-1. Ice-bath calibration of temperature sensors.

Incorrect

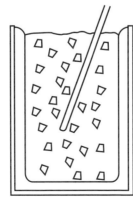

Correct

tip of the probe inside the fish (T_{COLD}), heating the probe and giving an erroneously high temperature reading.

Time Constant

When a temperature sensor experiences a step change in its surrounding temperature, as for example when it is thrust into a fish, it will take time for that reading to adjust (Figure A-3).

Engineers use the term "time constant" to describe the time it takes for about $^2/_3$ of the change to take place. For a very small thermocouple thrust into an ice-bath, this could be less than a second. But if the sensor is heavy and it is exposed to a medium in which heat transfer is fairly slow (like still air), the time constant could be many minutes. Watch out for this if you're trying to measure rapidly changing temperatures using a big sensor with a large time constant.

Figure A-2. Reduce conduction error.

(from Graham 1977)

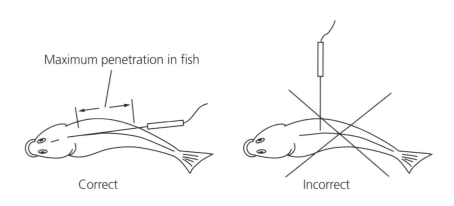

Insertion of thermometer in fish

Figure A-3. Time constant of a temperature transient.

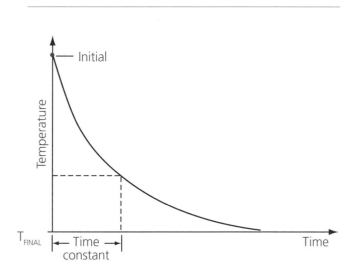

Lack of Steady State

Sometimes you're trying to catch the temperature of a product that you just pulled from the freezer. The sensor's time-constant will affect your accuracy, as described above. But you'll also need to think about the non-steady temperature of the product itself. When the just-frozen product is exposed to the warm air in the process room, it will begin to warm. Blocks could warm up at the rate of 1°F every minute or two. Smaller products such as fillets or crab sections will heat up much faster. We've measured core temperature of small fillets that increased 25°F within 10 minutes of leaving a spiral freezer.

FREEZING EQUIPMENT SUPPLIERS

Product Code

BL   Stationary Blast Freezer
BR   Brine Freezer
C   Cabinet
CO   Contact
CC   Cryogenics Using CO_2
CN   Cryogenics Using Liquid Nitrogen

F   Fluidized Bed
I   Immersion (LN)
PH   Horizontal Plate Freezer
PV   Vertical Plate Freezer
S   Spiral Freezer
SH   Shelf Freezer
T   Tunnel Freezer (Linear Belt Freezer)

Company	Products	Notes
Advanced Equipment, Inc. 2411 Vauxhall Place Richmond, BC Canada V6V 1Z5 (604) 276-8989 www.advancedfreezer.com	BL, CO, F, S, T	
Aerofreeze, Inc. 18394 Redmond Way Redmond, WA 98052 (425) 869-8889 www.aerofreeze.com	S, T	Manufacturing plant is in Canada
Air Liquide Industrial U.S.LP 2700 Post Oak Blvd. Houston, TX 77056-8229 (800) 820-2522 www.us.airliquide.com	C, CC, CN, I, T	Formerly Liquid Air Trade names: "Ultrafreeze," "Cryomix," "Crust flow," "Aligal" Supplies gas for food freezing Manufactures tunnel freezers; supplies immersion, batch/cabinet, and specialty freezers
Air Products and Chemicals, Inc. 7201 Hamilton Blvd Allentown, PA 18195-1501 (610) 481-5900 www.airproducts.com	C, CN, I, T	Trade names: "Cryo-Quick," "Freshline," "Cryo-tumbler"
BOC Gases 575 Mountain Ave. Murray Hill, NJ 07974 (908) 464-8100 www.boc.com	C, CC, CN, F, I, S, T	Formerly Airco Gases Trade name: "Cryomaster"
Carnitech US Inc. 1112 NW Leary Way Seattle, WA 98107-5133 (206) 781-1827 www.carnitech.com	BL, BR, S, T	Trade name: "HiFlow"

Freezing Equipment Suppliers continued

Company	Products	Notes
Cold Sea Refrigeration 758 Tillamuk Dr. LaConner, WA 98257 (360) 466-5850 (877) 265-3732	BL, BR, SH	
CSE - Cryogenic Systems Equip. 2363 W 136th St. Blue Island, IL 60406 (708) 385-4216 www.cryobrain.com	C, CC, CN, I, S, T	
Dole Refrigerating Co. 1420 Higgs Road Lewisburg, TN 37091 (800) 251-8990 www.doleref.com	PH, PV	Trade name: "Freze-cel"
DSI - Samifi Freezers Available through major U.S. refrigeration contractors www.dsi.as.com	PH, PV	Company based in Denmark and Italy Horizontal plate freezers advertised for low-temperature conditions using CO_2 refrigerant
Frigoscandia Equipment FMC Food Tech Chicago 200 East Randolf Chicago, IL 60601 (312) 861-6000 www.fmctechnlogies.com/FoodTech.aspx	BL, CO, F, S, T	A business of FMC Food Tech Freezing equipment is supplied by Frigoscandia and Northfield Trade names: "Nautica" (tunnel impingement freezer), "Advantec" (impingement freezer), "SuperCONTACT," "FloFREEZE"
G&F Systems 208 Babylon Turnpike Roosevelt, NY 11575 (516) 868-4923 www.gfsystems.com	S	

Company	Products	Notes
Gunthela Enterprise, Ltd. 962 Chatsworth Rd. Quaticum Beach, BC V9K 1V5 Canada (205) 752-4435 (877) 752-3311 www.gunthela.com	BR, C, SH	
IMS (Integrated Marine Systems) P.O. Box 2028 (775 Haines Place) Port Townsend, WA 99368 (360) 385-0077 (800) 999-0765 www.imspacific.com	BL, BR	Trade name: "Hydrochiller" Products include hatch-mounted blast freezers for onboard systems
Intec USA LLC 4319 South Alston Ave. Suite 105 Durham, NC 27713 (919) 433-0131 www.intecvrt.com	S, T	Company based in New Zealand
Jackstone Froster Ltd. Available through major U.S. refrigeration contractors www.jackstonefroster.com	PH, PV, SH	Company based in United Kingdom
Koach Engineering and Mfg. 8950 Glenoaks Blvd. Sun Valley, CA 91352-2059 (818) 768-0222 www.koachengineering.com	C, CC, CN, I, T	Trade name: "Cryomech" (cryomechanical tunnel freezer)
Linde Gas LLC 6055 Rockside Woods Blvd. Cleveland, OH 44131 (216) 573-7811 www.us.lindegas.com	C, CC, CN, S, T	Trade name: "Cryo-line"
Martin/Baron Inc. 5454 Second St. Irwindale, CA 91706-2060 (626) 960-5155 (800) 492-3765	C, CC, CN, T	

Freezing Equipment Suppliers continued

Company	Products	Notes
Pacific West Refrigeration 1647 Field Rd. Sechelt, BC V0N 3A1 Canada (604) 885-3499 (866) 885-3499	C, CO, SH	Trade name: "Sea Monster" (cabinet freezer)
Praxair, Inc. 39 Old Ridgebury Road Danbury, CT 06810 (800) 772-9247 www.praxair.com/food	CC, CN, I, S, T	Praxair acquired Liquid Carbonic in 1995 Trade names: "Coldfront," "Cryoshield" (CO_2), "Nitroshield" (LN) Freezing Headquarters in Wooster, Ohio: (800) 334-5242
Process Engr. and Fabrication 20 Hedge Lane Afton, VA 22920 (540) 456-8163 www.processengineeringinc.com	S	
RMF Freezers, Inc. 4417 East 119th St. Grandview, MO 64030 (816) 765-4101 www.rmfsteel.com	BL, S, T	Formerly Freezing Systems, Inc. Trade names: "Coldzone series" (crust freezer), "Cold- star series" (spirals), "Cold- wave series" (IQF tunnel)
Sandvik Process Systesm, LLC 21 Campus Rd. Totowa, NJ 07512 (973) 790-1600 www.processsystems.sandvik.com	CO, T	
Seattle Refrigeration and Mfg. Co. 1057 South Director St. Seattle, WA 98108 (206) 762-7740 (800) 228-8881 www.seafrig.com	BL, T	

Company	Products	Notes
Spiralsystems.com 11294 Coloma Rd. Rancho Cordova, CA 95670 (916) 852-0177 (800) 998-6111 www.spiralsystems.com	S	
Technicold Rich Beers Marine, Inc. 230 Southwest 27th St. Fort Lauderdale, FL 33315 (954) 764-6192 www.richbeersmarine.com	BR	Onboard freezing systems
Wescold *Seattle Division*: W.E. Stone P.O. Box 99185 (4220 22nd Ave. West) Seattle, WA 98199 (206) 284-5710 (800) 562-1945 *Portland Division*: P.O. Box 14250 (2112 SE 8th Ave) Portland, OR 97214 (503) 235-3193 (800) 547-2004	BR, PH	Trade names: "Century" (plate freezers), "Port-A-Chiller" (galvanized box chillers)
I.J. White 20 Executive Blvd. Farmington, NY 11735 (631) 293-2211 www.ijwhite.com	BL, S	

Index